Can Organisms Be Technology?

Edmund Russell

When environmental historians think about organisms at the intersection of environmental and technological history, they probably picture ways in which technology has harmed, first, wild species and, second, human health. These responses should not surprise us, for these topics have accounted for some of the most important works in the field. Historians of technology, on the other hand, might be more likely to call to mind ways in which human beings have used technology to remake the environment to serve human needs.[1] But can organisms themselves be technology? Some who have considered the question have answered in the negative.[2] This essay answers the question with an emphatic yes.

Resistance to seeing organisms as technology may arise from the dominant role the Industrial Revolution has played in the history of technology. Seeing technological history through the lens of that revolution has resulted in a focus on machinery, an acceptance of the idea that industrial technologies replaced nature, and a sense that only nonliving parts of nature can be technologies.[3] But those ideas merit questioning. All machinery is indeed technology, but not all technology need be machinery.[4] Human beings have used many tools throughout history, only some of them mechanical, and all tools are technologies. Human beings have never replaced nature with some wholly new technology; they have only transformed nature from one form into another, more useful form. And although many technologies deploy nonliving substances, there is no reason why some of our tools could not also be alive.[5]

Organisms become tools when human beings use them to serve human ends. Wild organisms in an untouched state of nature have not been human technologies, but many other organisms have become human instruments. In

Animals as technologies for gambling. Two hundred years ago, the English used bulls and bulldogs as tools for gambling. At bull baits, gamblers bet on whether the dog or the bull would get the better of the other. (From Henry Alken, *The National Sports of Great Britain* [New York: Appleton, 1903], 48; courtesy of Special Collections, University of Virginia)

some cases, such as in marshes that filter pollutants, humans use wild organisms without radically redesigning them, while in other cases they transform them to do work.[6] They redesign them just as they do nonliving technologies. Most domestic plants and animals have undergone such transformations. Those species are the product of artifice; they have become artifacts.[7]

Historians a hundred years from now will surely take it for granted that organisms can be technologies. They will live in a world populated by biotechnologies resulting from genetic engineering. The terms *biotechnology* and *genetic engineering* illustrate the synthesis of the natural (*bio*, meaning "life," and *gen*, meaning "born") and the artificial (*technology, engineering*) to create artifacts. The U.S. Patent Office, charged with patenting useful inventions, has already weighed in by issuing patents for genes and life forms. Much as we tend to see genetic engineering as carrying us into a brave new world (which it does), it marks but the latest episode in the development of biotechnologies. For some ten thousand years human beings have been transforming wild species into domestic versions to make agriculture possible. All settled societies, including ours, exist only on the sufferance of biotechnologies. It is important to recognize the long history of biotechnologies because, among other things, it provides us with a useful prehistory for understanding genetic engineering.[8]

Ways to Think about Organisms as Technology

One way to think about organisms as technology is to regard them as products.[9] As the accompanying tables illustrate, we use many types of organisms—mammals, birds, insects, plants, fungi, and microorganisms. And the types of products are wide ranging. Everything from food to fuel, edibles to entertainment, and construction materials to companions comes to us in the form of other species. We consume some organismal products, such as fruit, in their raw form. We process other products, such as silk for clothing. And we modify some organisms to produce more useful products. The ancestor of maize (teosinte), for example, produces only a few small kernels. By selecting bigger and better-tasting kernels over time, human beings have developed the familiar species with long, kernel-heavy ears, which forms an important part of economies around the world.

We even brand organismal products and protect those brands, or trademarks, through the same legal means we use for other products. Only apples grown in Washington can be sold as Washington apples, only potatoes grown in Idaho can be sold as Idaho potatoes, and only sparkling wines produced in Champagne can be sold as champagne. This may seem like common sense, but we have divested other organisms, such as Gouda cheese and Jersey cows, from geographical monopolies of names. Whether legally protected or not, those place names remind us that human beings redesigned organisms not just to produce better products but also to grow well in local environments.[10]

We have also striven for product standardization. One means has been to agree on classifications for products, such as grading criteria that indicate the putative quality of meat. Another has been inbreeding, mating close relatives to each other to minimize genetic variation, to increase the likelihood of offspring with predictable qualities. Yet another relies on closed breeding populations, in which only organisms registered as belonging to a group can be mated to each other. Dog breeders have been masters at these processes. Kennel clubs have developed "breed standards," or characteristics of the ideal individual of a given breed (were it ever produced), and then judged dogs according to those standards. Breeders have favored show winners in selecting sires and dams so that each generation tends to approach the standard more closely. Dog breeders have often mated parents to children, and siblings to each other, to preserve desirable qualities. And the purebred system forces

Table 1   PLANT BIOTECHNOLOGIES (INCLUDING LARGE FUNGI)

Service	Examples
Food for humans	*Many plant groups, including:* Cereals (oats, barley, rye, wheat, rice, maize) Pulses (beans, peas) Tubers (beets, turnips, carrots) Oil crops (safflower, rapeseed, coconut) Nuts (walnuts, palm nuts, figs, almonds) Fruits (pears, apples, oranges, grapes) Vegetables (cabbage, squash, lettuce, peppers) Spices (onions, garlic, coriander, cumin) Fungi (mushrooms, yeast)
Food for animals	Field crops (hay, soy, alfalfa, oats) Trees (leaves, acorns)
Fiber	Field crops (cotton, flax, hemp)
Fertilizer	Legumes Compost
Fuel	Trees (wood) Oil crops (biodiesel) Field crops (ethanol)
Drugs	Field crops (belladonna, digitalis, codeine, tobacco, alcohol, marijuana, opium, heroin) Fungi (psilocybin, penicillin)
Construction materials	Trees (lumber, floor coverings)
Manufacturing	Field crops (rice with human genes inserted to produce pharmaceuticals and compounds for infant formula)
Filtration of pollutants	Marshes (natural and artificial) Forests
Climate moderation	Trees (street and yard) Grass (insulating roofs)
Aesthetics	Grasses (lawns) Trees Shrubs Flowers Indoor plants

Table 2 ANIMAL BIOTECHNOLOGIES

Service	Examples
Food	*Many animal groups, including:* Mammals (cows, sheep, goats, pigs) Birds (chickens, turkeys, ostriches, pigeons) Fish (salmon, tilapia) Insects (honey, termite larvae) Mollusks (clams, oysters, snails)
Labor	Mammals for traction (horses, buffalo, bullocks, asses, reindeer, dogs, llamas, camels, elephants) Mammals for other jobs (elephants in logging, dogs in herding, cats in ratting, dogs for people with disabilities) Birds (passenger pigeons)
Companionship	Mammals (dogs, cats) Birds (parakeets) Reptiles (snakes, turtles) Fish (aquaria fish)
Leather, fur, feathers, fiber	Mammals (sheep, camels, llamas, cows, horses) Birds (chickens, geese) Insects (silkworms)
Pest control	Mammals (mongooses) Birds (ducks, geese) Insects (predators on other insects)
Construction materials	Mammals (horsehair in plaster) Insects (shellac, lacquer)
Wax	Insects (bees)
Oil and fat	Fish Mammals (cows, horses)
Jewels	Mollusks (pearl oysters)

Table 3 MICROSCOPIC BIOTECHNOLOGIES (BACTERIA, FUNGI)

Service	Examples
Food processing	Fungi (yeast for bread, beer, wine) Bacteria (yogurt, cheese)
Pathogens	Bacteria (pest control, biological weapons) Viruses (pest control, biological weapons) Fungi
Manufacturing	Bacteria (*E. coli* for making antibiotics)

breeders to use only registered animals as parents. The result has been remarkably successful. Even people with no interest in dogs can identify many breeds by sight, and *purebred* is a watchword for quality.[11]

Another way to think about organisms is as factories.[12] Plants and animals transform one kind of product into another, such as hay and water into milk and meat, atmospheric nitrogen into fixed nitrogen, and sun, water, and carbon dioxide into sugar. Plants and animals often process raw materials more efficiently than do machines created by human beings. We rely little on solar power harvested through manufactured collectors because the price per calorie is high. We rely instead on solar power harvested by plants to power our bodies and our economies. In some cases plants stored the sun's energy eons ago, and we now liberate that energy when we burn coal, oil, and natural gas. In other cases we rely on recent solar energy captured by plants, as when we eat bread made from last summer's wheat.[13]

Sometimes we have organized these minifactories into assembly lines. One example is the production of cheese. First we use grass to convert solar energy, carbon dioxide, and water into sugar and starches. Next we feed the grass to cows to convert its sugar and starch into meat (protein and fat) and milk (protein, fat, and lactose). Then we enlist microorganisms (molds, yeast, and bacteria) to transform milk protein into amino acids, fats into fatty acids, and lactose into other acids (such as lactic and acetic), and we call the result cheese.[14]

We also rely on organisms to be inventors. Organisms created many of the substances human beings have relied on throughout their history to survive and prosper, and it is hard to imagine that people would have imagined, much less come up with, ways to manufacture those materials. How would we have dreamed up a substance that increased its size on its own, delivered shade while being manufactured, was light yet strong, could float or sink as we wished, could be cut and joined with simple tools, could make objects ranging in size from tiny tools to mammoth Buddhist temples, and could easily be converted into heat and light? Yet wood is all of those things and more. We have created substances to mimic wood's many advantages (and reduce some of its disadvantages), but most of them are far less flexible. Steel studs used to build walls, for example, are strong and resist fire, but they do not reproduce themselves, they require specialized tools, they can be cut to length but not resized in cross section on job sites, and they cannot be burned for warmth.

Organisms also have served as workers.[15] In some cases they have enhanced or replaced human labor, as when herding dogs assist shepherds.

Such dogs have the added advantage of costing far less than human workers. In other cases organisms perform jobs that human beings cannot perform no matter how well trained. People lack the noses to sniff out bombs and drugs, but dogs accomplish that task easily.[16] In some cases organisms have served as technological alternatives to nonliving technology. Camels apparently superseded wheeled carts as transportation technology in the Middle East for centuries because they were better suited to the terrain.[17] Arctic peoples have tamed reindeer to serve as milk producers, draught animals, and beasts of burden.[18] Organisms have the added advantage of flexibility. If an organism performs less than optimally as a factory, inventor, or worker, reshape it. We have done so over and over again.

How Organisms Become Technologies

Many organisms do useful work for human beings without our modifying them much at all. Wild plants filter water pollutants from rivers before the water flows to the sea, transform the carbon dioxide we exhale into usable oxygen, protect stream resources from erosion by anchoring hillside soil, and do so many more things that we could spend a lifetime enumerating them. Those organisms and their services deserve attention from technological historians for the same reasons that ecological economists are now trying to quantify the goods and services that natural systems provide. One reason is intellectual: to understand any economy, we need to take into account the goods and services nature provides, as well as those that people provide. Another is practical: in standard economic methods, the loss of a free natural service is not counted as a cost, but markets require accurate pricing to function efficiently, so prices should incorporate the cost of lost common goods (such as marshes that filter pollutants).[19]

Here, though, I focus on organisms that human beings have intentionally transformed from their wild state into organisms doing human work. It is more accurate to think about a continuum of transformations rather than an absolute line between the wild and the artificial, for we have seen changes of varying degrees. To bring some order to the continuum, though, I shall group them into five categories: capture, taming, domestication, breeding, and genetic engineering.

Capture requires the least transformation of a wild organism into something useful. Human beings have long gathered plants and hunted animals for food, shelter, clothing, and other useful products. Bison on the American Great Plains, for example, furnished Native Americans with meat, grease, blankets, leather, and sinew. Today, machinery in factories produces synthetic versions of the last three products. In the nineteenth century some factories relied on bison leather for the belts that turned their machinery, illustrating the continued importance of animal technology even in the era of industrialization.[20]

Taming is training wild animals to be more docile and compliant than in the wild. The Asian logging industry long relied on wild elephants, captured and tamed, to do a variety of tasks. Elephants moved logs from place to place, towed heavy objects, and were ridden. They were especially useful in the period before good roads and railroads penetrated isolated areas.[21]

Taming changes the behavior of organisms; domestication changes their genetics as well as their behavior. In domestication, plants and animals breed in captivity and change their traits so as to become more useful to human beings. Domestic wheat produces bigger grains, domestic shrubs develop sweeter berries, and domestic pigs lay on more meat than their wild relatives. Judged by the size of their populations, many organisms prosper better under domestication than in the wild. The wild ancestor of the domestic silkworm has gone extinct, and millions more dogs than their ancestors, wolves, inhabit North America.[22]

We tend to think of taming as the first step toward domestication, but accident probably played a critical role in many cases. Take the development of the dog from the wolf. Human beings had domesticated no animals before the wolf, so it is hard to imagine that they had a template in mind for how to proceed. It is also hard to imagine that they looked at a wolf and planned to convert it into a sheepdog. A more likely scenario is that wolves became dogs through a series of small steps, most of them intended to produce nothing other than what was already evident.[23]

Wolves might have started the process by hanging around the outskirts of human encampments to scavenge bones and other trash as food. Human beings might have found the wolves useful as garbage collectors and as sentinels warning of approaching animals or other people. They might have started tossing excess food to the wolves after killing animals too large to consume before putrefaction. The wolves willing to come closest to human beings would have obtained more food than those far away, which would have enabled them to produce more offspring, which in turn would have made the camp followers less skittish over the years. Eventually some wolves might have been calm enough to enter the camp itself. They could have become pets and walking meat lockers for lean times. When choosing which wolf-dog to slaughter, the people would have killed the most obstreperous first, further tilting reproduction toward gentleness. Eventually the product looked and behaved differently enough from wolves that it received its own name (the Paleolithic equivalent of *dog*).[24] From there it would have been possible, accidentally or by design, to further differentiate the dog into breeds for various jobs.

Something similar probably happened with plants. Human foragers would have eaten the biggest, sweetest berries in a patch, transported their seeds back to camp in their guts, and deposited the seeds of those plants near camp when defecating. (Many seeds are designed to be dispersed by mammals.) Those seeds would have sprouted into shrubs near camp, where their biggest, sweetest berries were likely to be eaten, and so on over generations until a domestic berry evolved that looked and tasted different from its ancestors. At some point, as with dogs, people began deliberately breeding plants for traits and creating distinct varieties.

The next category is breeding, or conscious modification of plants and animals. Human beings have relied on two main methods. The first is culling, or simply eliminating less desirable individuals. Like the process of domestication, culling shifts the traits of a population over time without anyone worrying about which individuals mates with which. The second is selective breeding, which perpetuates desirable traits of the parents by forcing certain males and females to mate (via coupling in animals and pollination in plants). Observers have long noted that this method makes organisms into better tools to do human work. As N. S. Shaler put it in 1894, "Selection in the case of our horned cattle has within a few centuries converted the cows into mild-mannered and sedentary milk-making machines, and has deprived the bulls of the greater part of their ancient savage humor."[25]

The fifth category is genetic engineering, or moving individual genes from one organism to another using techniques from molecular biology. If the two organisms are of the same or closely related species, genetic engineering offers a faster, more precise way of doing what breeders have long done. But the novelty of genetic engineering lies in its ability to move genes across widely divergent taxonomic groups, such as between plants and animals. Scientists implant firefly genes in tobacco plants and frogs to make them glow, and rice plants in Missouri manufacture human proteins courtesy of inserted human genes.

Opponents of genetic engineering usually portray it as something radically new, which it is. Never before have we been able to move genes between plants and animals. Proponents of genetic engineering, on the other hand, portray it as but the latest phase in the production of biotechnologies, which it also is. For all of us who think history has something useful to say about the present and the future, it is essential that we recognize both the disjunction and the continuity between past and present. To understand the social forces that might push people to develop certain kinds of biotechnologies, we can look at past biotechnologies. For the most part, advocates of genetic engineering have emphasized its value in food production and medicine. Serving those sectors will surely motivate much of biotechnology development. But other forces, such as the desire to produce illegal drugs and biological weapons, have also driven biotechnology development in the past. We should be prepared to see at least the same range of motivations in the future that we have seen in the past.

In making accidental and intentional changes in other species, human beings have become evolutionary agents. Many of us tend to equate evolution with speciation, which makes it something nature did eons in the past. But evolution can be large or small, caused by any force (including human beings), and can occur at any time (including the present). We have shaped the evolution of all domestic species in fundamental ways, making us their partners in their development on the earth. Especially striking is our creation of organisms carrying genes from entirely different taxonomic groups. Through the organisms we modify, via whatever method, we have the potential to reshape wild animals in turn. One concern about farm-raised salmon, which carry genetically modified genes for rapid growth, is that they will escape and interbreed with wild salmon. Plants too might cross-pollinate with wild relatives.[26]

Historians of the environment and of technology (as this books shows, the two overlap) have recently incorporated attention to anthropogenic evolution under the banner of evolutionary history. Biologists use *evolutionary history* to refer to the ancestry of specific organisms. Now historians are using the term to mean a field or research program of history that examines the ways human beings have shaped the genetics of other species. In making organisms into tools, we have joined the evolutionary dance of history.[27]

Conclusion

Modern life would be impossible without anthropogenic evolution. If plants and animals had not been domesticated to make them suitable for agriculture, we would not be able to produce the agricultural surplus that makes urban societies possible. We would not have specialized occupations, most of the technologies we use would not exist, and these words would never have appeared on a page. To understand modern life, then, we have to recognize that shaping organisms into tools to do human work is the most consequential thing human beings have ever done. We will continue to do so as long as we and other organisms remain on the earth. We do not know what sort of world our changes will produce. We only know that the role of human beings in the ecology and evolution of other species has grown larger over time. Today our impact is accelerating. The changes we make in turn open new opportunities for human experience and perhaps close off others. In reshaping organisms to do human work, we have reshaped what it means to be human.[28]

Notes

1. The literature of these two fields is enormous. For a sampling, with an emphasis on the intersection of technology and the environment, see Donald Worster, *Dust Bowl: The Southern Plains in the 1930s* (Oxford: Oxford University Press, 1979); William Cronon, *Changes in the Land: Indians, Colonists, and the Ecology of New England* (New York: Hill & Wang, 1983); Carolyn Merchant, *The Death of Nature: Women, Ecology, and the Scientific Revolution* (San Francisco: Harper & Row, 1980); W. Bernard Carlson, *Technology in World History*, 7 vols. (New York: Oxford University Press, 2005); Martin V. Melosi, *Garbage in the Cities: Refuse, Reform, and the Environment, 1880-1980* (College Station: Texas A&M University Press, 1981); Jeffrey K. Stine, *Mixing the Waters: Environment, Politics, and the Building of the Tennessee-Tombigbee Waterway* (Akron, OH: University of Akron Press, 1993); Joel A. Tarr, *The Search for the Ultimate Sink: Urban Pollution in Historical Perspective* (Akron, OH: University of Akron Press, 1996); James C. Williams, *Energy and the Making of Modern California* (Akron, OH: University of Akron Press, 1997); Jeffrey K. Stine and Joel A. Tarr, "At the Intersection of Histories: Technology and the Environment," *Technology and Culture* 39 (October 1998): 601–40; Brian Black, *Petrolia: The Landscape of America's First Oil Boom* (Baltimore: Johns Hopkins University Press, 2000); Adam Rome, *The Bulldozer in the Countryside: Suburban Sprawl and the Rise of American Environmentalism* (New York: Cambridge University Press, 2001); Edmund Russell, "'Speaking of Annihilation': Mobilizing for War against Human and Insect Enemies, 1914–1945," *Journal of American History* 82 (March 1996): 1505–29; idem, "'Lost among the Parts per Billion': Ecological Protection at the United States Environmental Protection Agency, 1970–1993," *Environmental History* 2 (January 1997): 29–51; and idem, "The Strange Career of DDT: Experts, Federal Capacity, and 'Environmentalism' in World War II," *Technology and Culture* 40 (October 1999): 770–96.
2. See Envirotech, "Are Animals Technology?" 20–27 July 2001, and idem, "More Animals as Technology," 28 July–1 August 2001, both at http://www.udel.edu/History/gpetrick/envirotech (accessed 28 January 2002).
3. Another reason is probably the relationship that has developed between human beings and domesticated animals over time. Pet keeping—in which animals have given names, often live in the house, and never become dinner—encourages people to see animals more as sentient beings and less as tools.
4. Here *machinery* refers to nonliving physical objects, usually having moving parts, produced by human beings to do tasks and controlled through any of a variety of means.
5. The meaning of the term *technology* has changed over time. Originally it referred to the useful arts, that is, the ideas and techniques that enabled people to do useful things, or to the study of those ideas and techniques. Later the meaning shifted to emphasize the physi-

cal products of those arts, that is, objects themselves. I use *technology* primarily in the latter sense, of objects, though the technique that made many organisms so useful—breeding—qualifies as one of the most important of the useful arts. Eric Schatzberg, "Technik Comes to America: Changing Meanings of Technology before 1930," *Technology and Culture* 47 (July 2006): 486–512.

6. Here *work* means any tasks performed to serve the desires of human beings, including the provision of companionship.

7. See Ann Vileisis's essay "Are Tomatoes Natural?" in this volume.

8. On the technologies produced through genetic engineering, see Biotechnology Industry Organization, *Guide to Biotechnology* (Washington, DC, n.d.), http://www.bio.org/speeches/pubs/er/ (accessed 18 February 2007).

9. Here and below, I draw on ideas in Edmund Russell, "Introduction: The Garden in the Machine; Toward an Evolutionary History of Technology," in *Industrializing Organisms: Introducing Evolutionary History*, ed. Susan R. Schrepfer and Philip Scranton (New York: Routledge, 2004), 1–16; and idem, "Evolutionary History: Prospectus for a New Field," *Environmental History* 8 (April 2003): 204–28. For overviews, see Edward Hyams, *Animals in the Service of Man: 10,000 Years of Domestication* (London: J. M. Dent & Sons, 1972); H. Epstein, *The Origin of the Domestic Animals of Africa*, 2 vols. (New York: Africana, 1971); Frederick E. Zeuner, *A History of Domesticated Animals* (London: Hutchinson, 1963); S. Bokonyi, *History of Domestic Mammals in Central and Eastern Europe* (Budapest: Akademiai Kiado, 1974); Jack R. Harlan, *Crops and Man* (Madison, WI: American Society of Agronomy, 1975); B. Brouk, *Plants Consumed by Man* (London: Academic Press, 1975); and Maarten J. Chrispeels and David Sadava, *Plants, Food, and People* (San Francisco: W. H. Freeman, 1977).

10. There may or may not be genetic or other differences between products grown in different places. The branding itself can serve any of a variety of functions, such as quality control, marketing, and monopoly preservation. For a case study of branding, as well as the effect of consumption on the shaping of other organisms in places far from the consumers, see John Soluri, "Accounting for Taste: Export Bananas, Mass Markets, and Panama Disease," *Environmental History* 7 (July 2002): 386–410.

11. On the evolution of modern breeding methods and their effects, see William Boyd, "Making Meat: Science, Technology, and American Poultry Production," *Technology and Culture* 42 (October 2001): 631–64; Deborah Fitzgerald, *The Business of Breeding: Hybrid Corn in Illinois, 1890–1940* (Ithaca, NY: Cornell University Press, 1990); Jack Ralph Kloppenburg Jr., *First the Seed: The Political Economy of Plant Biotechnology, 1492–2000* (New York: Cambridge University Press, 1988); John H. Perkins, *Geopolitics and the Green Revolution: Wheat, Genes, and the Cold War* (New York: Oxford University Press, 1997); and Harriet Ritvo, *The Animal Estate: The English and Other Creatures in the Victorian Era* (Cambridge, MA: Harvard University Press, 1987).

12. Here I use *factory* to refer to physical objects (plants and animals) that transform products from one kind to another. One can see this usage either as a metaphor (organisms change products in a manner parallel to the way machinery does) or as an expansion of the meaning of *factory* from referring to machinery to referring to organisms as well.

13. For an environmental history that puts types of energy sources at its center, see I. G. Simmons, *An Environmental History of Great Britain: From 10,000 Years Ago to the Present* (Edinburgh: Edinburgh University Press, 2001).

14. Valerie Essex Cheke, *The Story of Cheese-Making in Britain* (London: Routledge & Kegan Paul, 1959).

15. Juliet Clutton-Brock, *A Natural History of Domesticated Animals*, 2d ed. (Cambridge: Cambridge University Press, 1999).
16. The horse is an excellent example of an animal used for labor. Joel A. Tarr, "A Note on the Horse as an Urban Power Source," *Journal of Urban History* 25 (March 1999): 434–48; Clay McShane and Joel A. Tarr, *The Horse in the City: Living Machines in the Nineteenth Century* (Baltimore: Johns Hopkins University Press, 2007).
17. Richard Bulliet, *The Camel and the Wheel* (New York: Columbia University Press, 1990).
18. Robert Paine, *Herds of the Tundra: A Portrait of Saami Reindeer Pastoralism* (Washington, DC: Smithsonian Institution Press, 1994).
19. In these cases the wetlands function like factories, producing clean water from dirty raw materials. On ecological economics more broadly, see Robert Costanza, *An Introduction to Ecological Economics* (Boca Raton, FL: St. Lucie, 1997).
20. Andrew C. Isenberg, *The Destruction of the Bison: An Environmental History, 1750–1920* (New York: Cambridge University Press, 2000).
21. J. G. B. Stopford, "Neglected Source of Labour in Africa," *Journal of the Royal African Society* 1 (July 1902): 444–51.
22. For the argument that organisms did better under domestication than they would have in the wild, see Stephen Budiansky, *The Covenant of the Wild* (New York: William Morrow, 1992); idem, *The Truth About Dogs: An Inquiry into the Ancestry, Social Conventions, Mental Habits, and Moral Fiber of Canis Familiaris* (New York: Viking, 2000); Michael Pollan, *The Botany of Desire: A Plant's-Eye View of the World* (New York: Random House, 2001); and Raymond P. Coppinger and Charles Kay Smith, "The Domestication of Evolution," *Environmental Conservation* 10 (Winter 1983): 283–92. On domesticates, see Clutton-Brock, *Natural History of Domesticated Animals*; Temple Grandin, ed., *Genetics and the Behavior of Domestic Animals* (London: Academic Press, 1998); and Harlan, *Crops and Man*.
23. Jared Diamond, *Guns, Germs, and Steel: The Fates of Human Societies* (New York: Norton, 1998). Subsequent paragraphs also draw on this source.
24. Adult dogs often bark, for example, while adult wolves do so rarely.
25. N. S. Shaler, "The Dog," *Scribner's Magazine* 15 (June 1894): 699 (quotation); Nicholas Russell, *Like Engend'ring Like: Heredity and Animal Breeding in Early Modern England* (Cambridge: Cambridge University Press, 1986). Darwin found the variation among animals and plants and the methods of selection used to affect traits both fascinating and useful in developing his theory of evolution through natural selection. Charles Darwin, *Variation of Animals and Plants under Domestication*, vol. 1 (1875; reprint, New York: New York University Press, 1988).
26. Darwin emphasized the similarity between natural and human selection in Charles Darwin, *On the Origin of Species* (1859; reprint, New York: New York University Press, 1988). His successors also looked to examples from human-caused evolution to develop the modern synthesis in evolutionary biology, which brought together evolutionary theory and population genetics. Theodosius Dobzhansky, *Genetics and the Origin of Species* (New York: Columbia University Press, 1937). For more on ways in which human beings accidentally caused the evolution of other species, in this case insects, see Edmund Russell, *War and Nature: Fighting Humans and Insects with Chemicals from World War I to Silent Spring* (Cambridge: Cambridge University Press, 2001); and Joseph E. Taylor, *Making Salmon: An Environmental History of the Northwest Fisheries Crisis* (Seattle: University of Washington Press, 1999).
27. Russell, "Evolutionary History"; Schrepfer and Scranton, *Industrializing Organisms*. Biological evolution can also undermine new nonliving technologies. S. B. Levy, *The Antibiotic*

Paradox: How Miracle Drugs Are Destroying the Miracle (New York: Plenum, 1992). On the increasing effect of human beings on evolution in general, see Stephen Palumbi, *The Evolution Explosion: How Humans Drive Rapid Evolutionary Change* (New York: Norton, 2001). Some historians of technology have looked to evolution as a metaphor for the way nonliving technology develops. George Basalla, *The Evolution of Technology* (Cambridge: Cambridge University Press, 1988).

28. On the effect of domestication on culture, see Richard Bulliet, *Hunters, Herders, and Hamburgers: The Past and Future of Human-Animal Relationships* (New York: Columbia University Press, 2005).

5

The Wild and the Tamed

As the Civil War approached its end, the citizens of the United States faced the prospect of an Indian war in the Great Plains. John Evans, the governor of the Colorado Territory, predicted in 1864 that the coming hostilities "will be the largest Indian war this country ever had, extending from Texas to the British line involving nearly all the wild tribes of the plains."[1] Contention over the control of natural resources was the root cause of the brewing conflict. In the southern plains, the nomads struggled to preserve their stewardship of the bison herds, while Euroamericans sought access to ranchlands and the gold mines of Colorado.[2] For the southern plains nomads, the conflict reached its nadir on November 29, 1864, when Colonel John Chivington led two companies of Colorado volunteer cavalry in an attack on noncombatant Southern Cheyennes at Sand Creek. Chivington and his men killed one hundred fifty Cheyennes; two-thirds of the dead were women and children. The Commissioner of Indian Affairs and a Congressional committee rebuked those involved in the massacre, but Chivington won broad support among Euroamericans in Colorado, including Governor Evans.

In the northern plains, the conflict between Indians and Euroamericans over the control of natural resources also raged. The dispute of the northern plains nomads with the United States centered on the Bozeman Trail leading from Fort Laramie to the gold mines of Montana. The trail and the emigrants on it cut directly through the Powder River valley, the nomads' best remaining hunting territory in the northern plains. The scattered attacks of the Indians on miners and soldiers turned to concerted warfare in 1866 after the army constructed three forts along the trail to protect emigrants. The United States fared poorly in the conflict. On December 21, 1866, near Fort Phil Kearney, one of the three new posts, a Sioux force lured eighty cavalrymen into an ambush and obliterated the entire unit. In the aftermath of this defeat, Generals William Sherman and Ulysses Grant urged Congress "to provide means and troops to carry on

---

[1] Quoted in Philip Weeks, *Farewell, My Nation: The American Indian and the United States, 1820–1890* (Arlington Heights, Ill.: Harlan Davidson, 1990), 102.

[2] See James Mooney, *Calendar History of the Kiowa Indians* (Washington, D.C.: Smithsonian Institution Press, 1979), 321–322.

formidable hostilities against the Indians, until all the Indians or all the whites on the great plains, and between the settlements on the Missouri and the Pacific, are exterminated."[3] Congress, weary of the human and fiscal costs of war, rather urged a policy of negotiation and created a Peace Commission composed of military and civilian leaders to conclude treaties with the plains nomads.

The Peace Commissioners first addressed the war in the southern plains. (See Map 5.1.) In October, 1867, the Peace Commission convened a meeting with over five thousand Comanches, Kiowas, Southern Cheyennes, and Southern Arapahos in the valley of the Medicine Lodge Creek. The agreements that the assembly produced – the Treaties of Medicine Lodge – purported to end the three years of hostilities between the southern plains nomads and Euroamericans. The treaties allowed Euroamericans to build roads and lay tracks to the Colorado mines, and indeed to settle north of the Arkansas River. The Indians secured the right to hunt south of the Arkansas "so long as the buffalo may range thereon in such numbers as to justify the chase." The treaties, moreover, forbade Euroamericans to settle south of the Arkansas.[4]

By establishing the Arkansas River as the boundary between Indians and Euroamericans, the Treaties of Medicine Lodge appeared to resolve the problems posed by conflicting land use practices. Although the treaties prohibited Euroamericans from settling in the nomads' hunting territory, they did not expressly forbid Euroamerican hunters from pursuing bison south of the Arkansas, except in the nomads' relatively small reservations between the Arkansas and Red rivers in Indian Territory. Indian signatories maintained, however, that the Peace Commissioners had made oral promises to keep Euroamerican bison hunters north of the Arkansas. Richard Bussell, a bison hunter who attended the negotiations, recalled that he and other bison hunters promised not to hunt south of the Arkansas. As Bussell noted, however, "the hunters paid no attention to the treaty."[5] The Treaties of Medicine Lodge, therefore, while establishing

---

[3] "Letter from the Secretary of the Interior, Communicating ... information in relation to the late massacre of United States troops by Indians at or near Fort Phil. Kearney, in Dakota Territory." S.ex.doc. 16, 39th Cong., 2nd Sess.

[4] See Charles Kappler, ed., *Indian Affairs: Laws and Treaties*, vol. 2 (Washington: Government Printing Office, 1904), 887–895, 977–989. The line of demarcation between Euroamerican settlement and Indian hunting territory did not correspond precisely to the Arkansas River but ran along the southern border of the state of Kansas to the juncture of the Cimarron River and Buffalo Creek, then north to the Arkansas River, then down the Arkansas to the state border.

[5] Richard Bussell, interview with J. Evetts Haley, 19 July 1926, Canadian, Texas; Bussell, "Buffalo Hunting," 27 December 1929, Panhandle-Plains Historical Society, Canyon, Texas. Bussell, recalling the Treaty of Medicine Lodge over fifty years after it had been signed, referred to General William S. Harney as "General Harker." He was perhaps confusing Harney with General Charles Harker. Bussell also thought that the conference had convened "in 1871 or 1872." Despite these lapses, he remembered the presence at the conference of the Cheyenne leader Black Kettle, a treaty signatory. The bison hunter Mark Huselby also recalled that "the whites had made an agreement to stay back, but they had begun to encroach upon the plains, which were the Indians' hunting grounds." Huselby, interview with Haley, 18 June 1925, Mobeetie, Texas, Panhandle-Plains Historical Society.

the Arkansas River as the informal boundary between Indian and Euroamerican territory, did not guarantee to southern nomads the exclusive right to hunt bison south of the Arkansas. After 1870, Euroamerican hide hunters poured into the region and within a decade had nearly exterminated the herds.

After the conclusion of the negotiations at Medicine Lodge, the members of the Peace Commission traveled to Fort Laramie on the north branch of the Platte River, hoping to negotiate an end to the war between the United States and the Sioux, Northern Cheyennes, and Northern Arapahos. In his winter count, the Sioux calendarist Lone Dog memorialized the 1868 negotiations – which he termed "many flags were given" – as the principal event of the year.[6] Indeed, the conferences were critical to the northern plains nomads, who hoped to capitalize on their recent military successes against the United States and secure lasting control over their remaining hunting territory.

Like the agreements at Medicine Lodge, the Treaty of Fort Laramie of April, 1868, appeared to end the dispute over resources. In return for cession of the territory south of the Platte River by the northern plains nomads and the promise not to impede the construction of railroads outside of their hunting area, the Peace Commissioners agreed to abandon the forts in the Powder River valley and to close the Bozeman Trail. Moreover, the treaty explicitly guaranteed to the northern plains nomads their undisturbed use of the hunting territory west of the Missouri, north of the Platte, and east of the Bighorn Mountains. "So long as the buffalo may range thereon in such numbers as to justify the chase," the Treaty forbade Euroamericans to settle or to pass through the nomads' hunting territory.[7]

Yet just as the Treaties of Medicine Lodge had only appeared to protect the Indians' hunting territory, the Treaty of Fort Laramie was an imperfect guarantee of the nomads' undisturbed use of the bison herds. The seemingly straightforward treaty was only one part of the contradictory regulation of resource use in the plains. The agreement banned Euroamerican hunters from the Indians' territory, but the law simultaneously allowed scientific exploration, land grants, and mining claims – the legal framework of Euroamerican settlement – to go forward.[8] In 1864, the federal government had set aside land in the Dakota and Montana territories for the construction of the Northern Pacific Railroad. The Treaty of Fort Laramie did not specify a northern boundary of the Indians' hunting territory, virtually ensuring conflict in 1873 when the Northern Pacific, building westward from Duluth, Minnesota, reached the Missouri River. Moreover, in 1874, Lieutenant-Colonel George Armstrong Custer led a scientific

[6] Garrick Mallery, "Pictographs of the North American Indians," *Fourth Annual Report of the Bureau of Ethnology, 1882–83* (Washington, D.C.: Government Printing Office, 1886), 125.
[7] Kappler, *Indian Affairs*, vol. 2, 998–1005.
[8] Hendrik Hartog pointed out to me the problem of multiple sources of legal authority in the nineteenth-century United States. See his article, "Pigs and Positivism," *Wisconsin Law Review* (1985), 899–935.

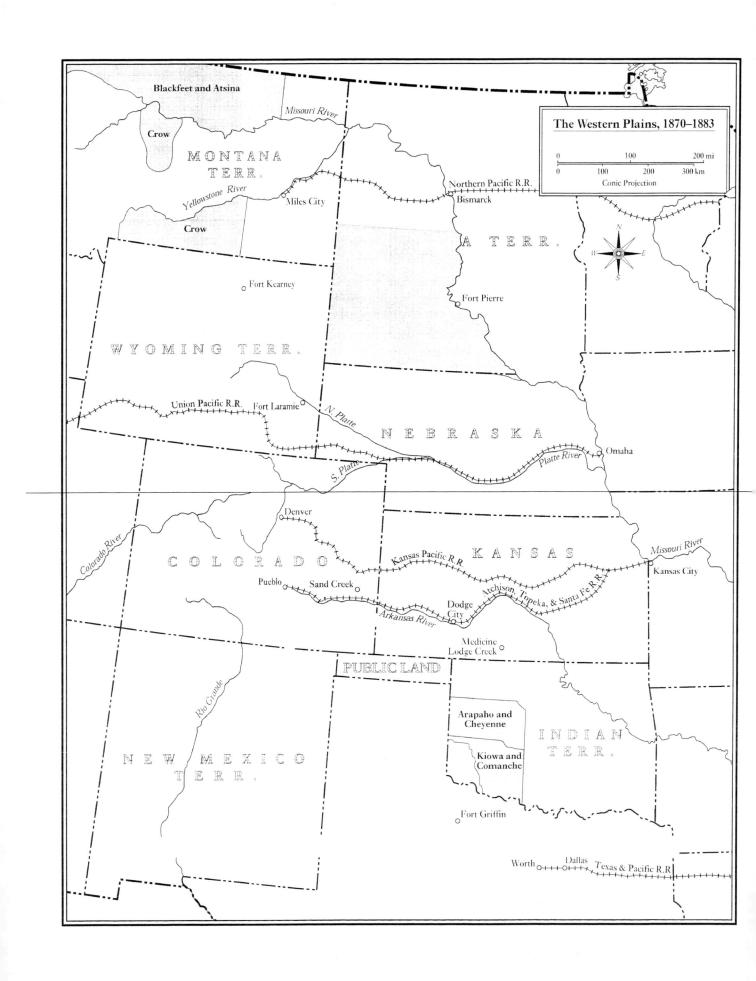

expedition to the Black Hills in the center of the area reserved for the Sioux. Custer's report of an abundance of gold in the Black Hills sparked a rush of Euroamerican miners into the Sioux territory. They were followed in the 1880s by Euroamerican hide hunters.

Above all, the agreements of Medicine Lodge and Fort Laramie were ephemeral because the United States' recognition of autonomous Indian hunting territories was due to expire with the bison. By 1867, both the Indian and Euroamerican signatories to the treaties knew that the bison population was declining rapidly. The diminution of the bison certainly did not escape the notice of General Sherman, a member of the Peace Commission. He wrote to his brother, John Sherman, a United States Senator from Ohio, on June 17, 1868, that "the commission for present peace had to concede a right to hunt buffaloes as long as they last, and this may lead to collisions, but it will not be long before all the buffaloes are extinct near and between the railroads."[9] The language of the treaties Sherman helped to negotiate was temporizing: the Indians "yet reserve the right to hunt ... so long as the buffalo range ... in such numbers to justify the chase." In 1870, Jacob Cox, the Secretary of the Interior, wrote somewhat prematurely that the provisions of the treaties reserving hunting territory to the nomads were moot because the Indians were already unable to subsist on the few remaining bison. "The building of the Union Pacific Railroad has driven the buffalo from their former hunting grounds," he wrote, so "that it was impracticable for the Indians to rely upon this natural supply of food, clothing, and shelter."[10] Sherman and Cox were ultimately proved correct; by 1883, the bison was nearly extinct and the plains nomads submitted to the reservation system.

The historian David D. Smits has argued that the United States Army was primarily responsible for the destruction of the bison. Certainly Sherman, the commander of the Division of the Missouri, suggested more than once that the eradication of the bison would force the plains nomads to reservations. The strategy, if indeed it was Sherman's aim to destroy the bison, was reminiscent of his destruction of Confederate resources during the Civil War. Sherman burned a path through Georgia and South Carolina in 1864–65, while General Philip Sheridan simultaneously conducted a similar campaign in the Shenandoah Valley in Virginia.[11] In the West in the 1860s and 1870s, the Army, when frustrated by its inability to defeat Indians in battle, sometimes resorted to the destruction of their animal resources. In 1864, Kit Carson, a breveted brigadier general in the First New Mexico Volunteer Infantry, ordered the destruction of Navajo sheep

---

[9] Rachel Thorndike Sherman, ed., *The Sherman Letters: Correspondence Between General and Senator Sherman from 1837 to 1891* (New York: Scribner's, 1894), 320.

[10] Jacob D. Cox, *Annual Report of the Secretary of the Interior*, 41st Cong., 1st Sess., 1870, H.exdoc. 1 (Serial 1449), vi.

[11] James M. McPherson, *Ordeal by Fire: The Civil War and Reconstruction*, 2d ed. (New York: McGraw-Hill, 1992), 443–446, 459–463; McPherson, *Battle Cry of Freedom: The Civil War Era* (New York: Ballantine, 1988), 774–806.

in order to bring the Indians to submission. In 1868, following his attack on a winter camp of Southern Cheyennes on the Washita River, Custer ordered his men to shoot eight hundred horses belonging to the Indians. Colonel Eugene Baker ordered three hundred Blackfeet horses killed following his attack on a Piegan winter camp on the Marias River in 1870.

In the 1870s, Smits argued, soldiers routinely led wasteful bison hunts.[12] Soldiers in the plains, however, were too few and too otherwise occupied to kill enough bison to make a difference to the Indians' subsistence. Their organized hunts killed only a few thousand bison in the 1860s and 1870s. Particularly in the southern plains, the Euroamerican slaughter of the bison in the 1870s was not primarily the work of soldiers, but of civilian hunters. The army declined to enforce treaty provisions banning the hunters from Indian territory and provided the civilian hunters with protection and sometimes ammunition. In other words, Sherman commended the hunters but he did not command them. Although the hunters served Sherman's purposes, they did not come to the plains at his behest but rather to satisfy an industrial economy's appetite for natural resources. The expansive American economy, which brought to the plains not only hide hunters but domestic cattle to replace the slaughtered bison, was primarily responsible for the destruction of the 1870s and early 1880s. Economic forces thus spared the army from fighting the extensive Indian war that Governor Evans had predicted in 1864.

Other historians have attributed the destruction of the bison not to uniformed or civilian hunters but primarily to environmental factors. They have pointed to the impact of drought, the predation of wolves, domestic livestock's pressure on plains forage, and exotic bovine diseases as the likeliest causes of the collapse of the bison population in the 1870s and early 1880s.[13] Environmental factors certainly contributed to the demise of the bison, but the decisive role of the environment is far easier to assert than to prove. Neither the predation of wolves nor the competition of other grazers in the 1870s and 1880s was sufficient to cause the destruction of the bison. Drought had periodically afflicted the plains for centuries, causing the bison population to fluctuate considerably, but the species had always recovered. Environmental factors are important when considered in *combination* with anthropogenic causes of the bison's decline, particularly the unprecedented, large-scale onslaught by Euroamerican commercial hide hunters.

---

[12] David D. Smits, "The Frontier Army and the Destruction of the Buffalo: 1865–1883," *Western Historical Quarterly*, 25 (Autumn 1994), 312–338.

[13] Rudolph W. Koucky, "The Buffalo Disaster of 1882," *North Dakota History*, 50 (Winter 1983), 23–30; Dan Flores, "Bison Ecology and Bison Diplomacy: The Southern Plains from 1800 to 1850," *Journal of American History*, 78 (September 1991), 465–485; Richard White, "Animals and Enterprise," in Clyde Milner, et al., eds., *The Oxford History of the American West* (New York: Oxford, 1994), 247–249; Elliott West, *The Way to the West: Essays on the Central Plains* (Albuquerque: University of New Mexico Press, 1995), 72–83.

If the natural environment contributed to the bison's demise, so too did the cultural environment. The army was the forward arm of federal government policy in the plains, which in turn reflected dominant cultural notions of the rightness of the Euroamerican conquest of the West. Most soldiers, emigrants, hunters, and ranchers in the plains shared a belief in the inevitability of Euroamerican triumph over Indians and the wilderness. They took the droughts, blizzards, and grass fires in the 1870s and 1880s that contributed to the destruction of the bison as evidence of the providential extinction of the herds. The belief that domestic livestock were destined to replace the bison sustained the hide hunters' destructive harvest. Some Euroamericans in the 1870s drew on opposing cultural traditions and argued for a policy of humanity toward both Indians and the bison. They argued that the bison and their primary users, the nomads, were already tame. Yet, the competing notion of the bison and the Indians as wild and therefore inferior to domestic cattle and their Euroamerican masters prevailed. Federal authorities institutionalized these cultural notions and channeled the transforming forces of the American industrial economy into the bison range. Together with inherent environmental pressures, those forces pushed the bison to the brink of extinction.

I

A spasm of industrial expansion was the primary cause of the bison's near-extinction in the 1870s and early 1880s. During this period, Euroamericans hunted bison to satisfy an increasing demand for hides in industrializing America. Leather belts were the sinews of nineteenth-century industrial production: mills relied on heavy leather belting to animate their machinery.[14] In 1850, leather manufacturing, based largely in New York and Pennsylvania, was the fifth-largest industry in the United States, after lumber, flour, boots and shoes, and blacksmithing. The tanning industry grew in value in the second half of the nineteenth century. The value of leather belting produced in the United States rose from over $6,500,000 in 1880 to over $8,600,000 in 1890, an increase of 32 percent.[15]

At the same time, however, the number of tanneries in the United States steadily declined, from over five thousand in 1860 to fewer than two thousand in 1890, as the industry was increasingly concentrated in fewer hands. The consolidation of smaller firms into larger conglomerates, to reduce the costs of

---

[14] Lucius F. Ellsworth, *Craft to National Industry in the Nineteenth Century: A Case Study of the Transformation of the New York State Tanning Industry* (New York: Arno Press, 1975), 193.

[15] See *Report on the Manufactures of the United States at the Tenth Census* (June 1, 1880). Misc. Doc. 42, 47th Cong., 2nd Sess. (Serial 2130), 9; *Report of the Manufacturing Industries in the United States at the Eleventh Census: 1890*, Misc. Doc. 340, 52nd Cong., 1st Sess., (Serial 3024), 126–127.

production, was typical of late nineteenth-century American business. In the tanning industry, however, consolidation was a symptom of crisis. Other industries lowered their production costs through increased mechanization, but tanning remained a largely labor-intensive process. While labor costs remained relatively high, the price of hides, the raw material of tanning, rose. In 1850, Adirondack tanneries paid an average of $2.20 per hide. By 1870, the average price had nearly doubled. Because the demands of the tanning industry outstripped the domestic supply, 70 percent of the hides tanned in the United States in 1850 were imported from Latin America, primarily from the Argentine pampas.[16] Continuing demand, as well as the high cost of cowhide produced in the United States, prompted the industry's interest in the bison as a relatively low-cost source of hides.

Exploitation of the bison required a heavy investment of industrial capital, energy, and knowledge. Before bison hides could become a viable commodity, tanners had to perfect the chemical process of transforming them into leather. Tanners began experimenting on bison hides in 1870. They eventually perfected a process that relied on soaking the hides in a strong lime solution.[17] The result was a porous but highly elastic leather ideally suited to industrial belting. Other technological innovations facilitated industrial America's absorption of the bison. The early single-shot rifles were too light, inaccurate, and difficult to reload to permit large-scale commercial bison hunting. Toward the end of the Civil War, however, American munitions makers developed accurate, large-bore rifles. In the late 1860s, bison hunters used the Springfield Army rifle, a .50-caliber weapon that fired with the power of seventy grains of powder. In 1872, the Sharp Rifle Manufacturing Company developed a still more powerful weapon that became the most common rifle among the bison hunters. The Sharp "big fifty" fired slugs weighing up to one pound with the power of ninety grains of powder, allowing hunters to fire accurate and deadly shots from several hundred yards away.[18] The extension of railroads to the plains allowed easy shipment to the East. The Union Pacific, which joined the Central Pacific in Utah in 1869, traversed the plains through Nebraska; the Kansas Pacific reached Denver in 1870; and the Atchison, Topeka, and Santa Fe reached Dodge City, in western Kansas, in 1872. The financial panic of 1873 stalled railroad expansion into the northern and extreme southern plains; the Texas and Pacific had reached Dallas and the Northern Pacific had extended to Bismarck, Dakota Territory, when the panic and the ensuing economic depression began. Construction on the Texas and

---

[16] Barbara McMartin, *Hides, Hemlocks, and Adirondack History: How the Tanning Industry Influenced the Region's Growth* (Utica, N.Y.: North Country Books, 1992), 8, 27, 88, 108; Victor S. Clark, *History of Manufactures in the United States, 1860–1914* (Washington, D.C.: Carnegie Institute, 1928), 465–466.

[17] Henry R. Proctor, *A Text-Book of Tanning: A Treatise on the Conversion of Skins into Leather, Both Practical and Theoretical* (New York: E. & F.N. Spon, 1885), 144.

[18] Lewis T. Nordyke, "King of the Buffalo Slayers," *Farm and Ranch* (April 1939), 7.

Pacific resumed in 1878 and had reached nearly to El Paso by 1881. The Northern Pacific resumed construction in 1879 and reached Glendive, Montana Territory, in 1880.[19] In sum, the hunting of the bison in the 1870s and early 1880s was unquestionably the work of an industrial society. The western plains became a remote extension of the global industrial economy and an object of its demand for natural resources.

The destruction of the bison in the plains to fuel the demand for hides was part of a broad pattern of environmental degradation in industrializing America. The tanning industry consumed not only animal hides but trees and lime in enormous quantities; it returned unused portions of those resources to local environments in the form of pollution. The tanning industry initially concentrated in northeastern Pennsylvania and the Adirondack region of New York, amidst dense forests of Eastern hemlock (*Tsuga canadensis*), because the bark of hemlock is rich in tannin, the chemical essential to the production of leather. Heavy harvesting of Eastern hemlock rapidly depleted the species, whose pace of reproduction was not suited to industrial consumption. The species grows slowly, reaching maturity in about three hundred years. Following cutting in the nineteenth century, the second generation faltered because hemlock seedlings cannot endure the strong sunlight in cut-over areas. The industry's unsustainable consumption of tanbark had depleted accessible supplies of hemlock in the Adirondacks by the mid-1880s and a few decades later in Pennsylvania. The *New York Times* attributed the 1883 failure of Shaw and Brothers, the largest tanning company in the United States, to the firm's exhaustion of local tanbark supplies.[20] Deforestation had widespread environmental as well as economic consequences, ranging from the destruction of local wildlife habitats to increased erosion and runoff of soil into streams.[21] Tanneries were also a leading contributor to the noxious condition of rivers in the industrializing Northeast. Tanneries discharged large amounts of organic pollution, including particles of lime and animal hair and flesh, into rivers. Significant quantities of organic pollution, particularly in combination with inorganic pollutants from textile and paper mills, overwhelmed the capacity of aerobic bacteria to break down contaminants in many rivers in the Northeast.[22] Late nineteenth-century tanneries were thus an environmental malignancy that destroyed bison, razed forests, and fouled rivers.

---

[19] Lloyd J. Mercer, *Railroads and Land Grant Policy: A Study in Government Intervention* (New York: Academic Press, 1982), 44–54.

[20] *New York Times* (September 16, 1892), quoted in McMartin, *Hides, Hemlocks*, 119; Gordon G. Whitney, *From Coastal Wilderness to Fruited Plain: A History of Environmental Change in Temperate North America, 1500 to the Present* (New York: Cambridge University Press, 1994), 186–188.

[21] George Perkins Marsh, *Man and Nature, or, Physical Geography as Modified by Human Action*, ed. David Lowenthal (Cambridge, Mass.: Harvard University Press, 1965), 186–189.

[22] Theodore Steinberg, *Nature Incorporated: Industrialization and the Waters of New England* (New York: Cambridge University Press, 1991), 206–209.

Although the industrial system that brought Euroamerican hide hunters to the plains wrought extensive environmental change, Euroamerican hide hunters were not able to shape the environment to their liking completely; they also adapted themselves to the constraints of the western plains. Because the bison spent most of the year fragmented into small foraging groups, Euroamericans, like bands of nomads, dispersed in search of bison in small hunting outfits often based on kinship ties. A typical hunting outfit consisted of a shooter, a cook, and three or four skinners. John Cook signed on to serve in such an outfit in 1870. Charlie Hart, a survivor of the Confederate prisoner-of-war camp at Andersonville, did the shooting. Cook, Warren Dockum, Cyrus Reed, and Reed's "green, gawky" teen-aged brother-in-law Frank Williamson skinned the animals. A man known to Cook only as Hadley drove the team, cooked the outfit's meals, and managed the camp.[23] Such outfits staked out their camps along the rivers where the bison sought water. In the early 1870s, outfits were camped every few miles along the Smoky Hill, Arkansas, Cimarron, Canadian, Washita, and Red rivers.

Although they seemingly overcrowded the bison's range, dispersal was essential to the hunters' success, just as it was to nomadic Indian hunters. The Euroamericans were generally careful not to encroach on each other's hunting grounds. Colonel Richard Irving Dodge wrote, "there are unwritten regulations, recognized as laws, giving to each hunter certain rights of discovery and occupancy."[24] Like the nomads who battled for the control of hunting territory, hunters did not secure those rights without conflict. The skinner S. P. Merry remembered that "some of the buffalo hunters tried to act hoggish and claimed a certain range for their hunting ground, but they could not do it."[25] Disputes over discoveries of hunting grounds likely represented the exercise rather than the absence of informal regulations. Such arrangements were typical of Euroamerican resource exploitation in the nineteenth-century West. Squatters who moved into public lands and miners who established claims also improvised their regulations.[26]

Although they preferred to be called "buffalo runners," most hide hunters did not pursue bison from horseback like the plains nomads. Instead, they employed a method known as the "still hunt." At dawn, a hunter slipped up to the downwind

---

[23] John R. Cook, *The Border and the Buffalo: An Untold Story of the Southwestern Plains*, ed., Milo M. Quaife (Chicago: Lakeside, 1938), 159–160.
[24] Richard I. Dodge, *The Plains of North America and Their Inhabitants* (Newark: University of Delaware Press, 1989), 151. See also Frank Mayer and Charles B. Roth, *The Buffalo Harvest*, (Denver: Sage, 1958) 47; H. B. Lovett, interview with L. F. Sheffy, 23 June 1934, Pampa, Texas; and J. W. Woody, interview with Haley, 19 October 1926, Panhandle-Plains Historical Society.
[25] Merry, interview with Haley, Amarillo, Texas, 21 August 1926, Panhandle-Plains Historical Society.
[26] See Malcolm Rohrbough, *Days of Gold: The California Gold Rush and the American Nation* (Berkeley: University of California Press, 1997), 15–16; Arthur F. McEvoy, *The Fisherman's Problem: Ecology and Law in the California Fisheries, 1850–1980* (New York: Cambridge University Press, 1986), 95.

side of a group of bison and concealed himself several hundred yards away in a wallow or behind the rise of a bluff. Using a telescopic sight to scan the herd, he located the lead cow of the group and killed her with his first shot, which he aimed below the animal's shoulder to strike her lungs and the bottom of her heart. The remaining animals could hear the report of the rifle but neither see nor smell their enemy. Nor could they rely on their leader to alert them to the danger. As she collapsed, dropping first to one foreleg, then the other, the hunter picked off several more of the bewildered animals, taking care to kill the wariest bison before they smelled blood and led the herd out of range. Such a "stand" varied in productivity, from as few as ten to as many as fifty hides in one morning's work. The hunter's labors ended before noon, when the skinners took over. Ordinarily, they removed hides until sundown.[27]

Like the Indians who brought their bison robes to trading posts such as Bent's Fort, Fort Union, and Fort Pierre, many Euroamerican hunters hauled their hides to centers of trade. Dodge City, Kansas, was initially the capital of the hide trade. The Santa Fe railroad had hardly reached the city in September, 1872, before "the streets of Dodge were lined with wagons, bringing in hides and meat and getting supplies from early morning to late at night," recalled one of the town's early residents. In the winter of 1872–73, the largest hide dealers in Dodge City, Robert Wright and Charles Rath, shipped over two hundred thousand hides on the Santa Fe; Wright estimated that the other dealers in town shipped another two hundred thousand. The editor of a Dodge City newspaper wrote in 1877 that in Wright and Rath's warehouse, "it is no uncommon thing to find from sixty to eighty thousand buffalo robes and hides."[28] (See Figure 5.1.)

The impact of the hide hunters on the herds in the southern plains was devastating. In 1872, Colonel Dodge wrote that the area near Dodge City was thick with bison. By the fall of 1873, however, "where there were myriads of buffalo the year before, there were now myriads of carcasses. The air was foul with sickening stench, and the vast plain which only a short twelve months before teemed with animal life, was a dead, solitary putrid desert." Even those bison who avoided bullets felt the impact of the hunters. Dodge wrote that "Every drink of water, every mouthful of grass is at the expense of life, and the miserable animals continually harassed, are driven into localities far from their haunts, anywhere to avoid the unceasing pursuit."[29] George Lemmon, a Nebraska rancher, wrote that after 1875, bison rendered no tallow and only stringy meat, "as hide hunters

---

[27] For descriptions of still hunting, see Homer W. Wheeler, *Buffalo Days: Forty Years in the Old West; the Personal Narrative of a Cattleman, Indian Fighter, and Army Officer* (Indianapolis: Bobbs-Merrill, 1923), 80–81; Cook, *The Border and the Buffalo*, 168–176; Dodge, *Plains of North America*, 151–153. See also T. S. Bugher, interview with Haley, 17–18 July 1925; J. W. Woody, interview with Haley, 19 October 1926, Panhandle-Plains Historical Society.

[28] Robert M. Wright, *Dodge City, The Cowboy Capital* (Wichita, 1913), 75; Ford County *Globe*, quoted in Wright, 158.

[29] Dodge, *Plains of North America*, 150–151.

Figure 5.1. Charles Rath and Robert Wright's Dodge City hide yard, 1874. Courtesy of the Kansas State Historical Society

had the few remaining animals almost run to death."[30] Like those of some other mammals and birds such as the passenger pigeon – millions of which market hunters in eastern North America slaughtered in the 1870s – the bison's communal breeding system relied on the assembly of large groups in the summer. Reproductive success likely declined with group size in the 1870s, as unceasing predation prevented the congregation of the herds in the rutting season, upsetting the bison's patterns of migration and reproduction and thus inhibiting a recovery of the bison population.[31]

How many bison did the hide hunters slaughter in the 1870s? The productivity of the outfits varied. W. S. Glenn, who hunted bison in Texas in 1876–77, recalled that "a remarkably good hunter would kill seventy-five to a hundred in a day, an average hunter about fifty, and a common one twenty-five, some hardly enough to run a camp. It was just like in any other business. A good skinner would skin from sixty to seventy-five, an average man from thirty to forty, and a common one from fifteen to twenty-five."[32] Charlie Hart's outfit stacked up 2,003 hides in six weeks, an average of about 50 hides a day.[33] Bat Masterson's crew, which was camped close to Fort Dodge and, unlike most outfits, took the time to take the meat from the carcasses, averaged only 15 hides a day.[34]

Colonel Dodge conservatively estimated the number of hides shipped on the Santa Fe, Kansas Pacific, and Union Pacific railroads between 1872 and 1874 to be 1,378,359. His figures were largely guesswork because only the Santa Fe provided him with a statement of the number of hides shipped. Dodge assumed that the other two lines shipped a number of hides equal to the Santa Fe's 459,453. Yet the Santa Fe's accounting of the hides it shipped in this period – even if accurate – was no reflection of the actual number of bison killed. Dodge noted that in the early years of the slaughter hunters were especially wasteful. Poor hunters wounded two or three bison for every one they killed; the crippled animals later fell to wolves. Skinners ruined hides as they flayed them, or failed to stretch and stake them to dry properly. When skinners failed to finish their work before sundown, wolves had often torn the unskinned carcasses to pieces by the next morning. Dodge estimated that in 1872 every hide shipped to market represented five dead bison. As they learned their craft, the outfits became

---

[30] Lemmon Papers, State Historical Society of Nebraska, Lincoln.
[31] Douglas B. Bamforth, "Historical Documents and Bison Ecology on the Great Plains," *Plains Anthropologist*, 32 (February 1987), 1–16. For the passenger pigeon, see Jenny Price, "Flight Maps: Encounters with Nature in Modern America" (Ph.D. dissertation, Yale University, 1998); H. Ronald Pulliam and Thomas Caraco, "Living in Groups: Is There an Optimal Group Size?" in J. R. Krebs and N. B. Davis. eds., *Behavioural Ecology: An Evolutionary Approach*, 2d ed. (Sunderland, Mass.: Sinauer, 1984), 134.
[32] Rex W. Strickland, ed., "The Recollections of W. S. Glenn, Buffalo Hunter," *Panhandle-Plains Historical Review*, 22 (1949), 25–31.
[33] Cook, *The Border and the Buffalo*, 171.
[34] Henry H. Raymond, "Diary of a Dodge City Buffalo Hunter, 1872–73" ed., Joseph W. Snell, *Kansas Historical Quarterly*, 31 (Winter 1965): 350–351.

more efficient. In 1873 every hide represented only two dead bison. Altogether, Dodge estimated that hide hunters slaughtered 3,158,730 bison between 1872 and 1874, an average of just over one million animals each year.[35] The Euroamerican hunters resembled the Indian nomads in many ways; small groups dispersed over a broad range, relying on the bison's gregariousness, and brought their hides to centers of trade. Yet the destructive effect of the hide hunters was significantly greater than that of the more numerous nomads.

If the hide hunters' estimates of their own productivity were correct, the impact of commercial hide hunting was still more devastating than Dodge guessed. Glenn estimated that a "common" hunter killed between fifteen and twenty-five bison a day.[36] The Kansas hunter Frank Mayer set twenty-fives hides as his "regular quota."[37] An outfit that accumulated twenty-five hides a day could amass 2,250 hides in three months. Such a figure was not unreasonable. Charlie Hart and his outfit accumulated 3,361 hides in three months.[38] In one season, a hide hunter known as "Kentuck" brought in 3,700 hides.[39] J. W. Woody claimed to have killed 3,200 in the winter of 1876–77.[40] Thomas Linton of Troy, Kansas, killed 3,000 in 1872. Zack Light killed 2,300 in the winter of 1873–74. Joe McCombs killed 4,900 in Texas in the winter of 1877–78. William "Doc" Carver killed 5,700 in 1875. Orlando Brown killed 5,855 bison during a two-month span in 1876 – indeed, the nearly constant report of his .50-caliber rifle during that time rendered him deaf in one ear.[41] A competent hunter could have a single productive season – or inflate his estimates to appear more successful before his peers – but more common perhaps was Josiah Wright Mooar's tally. He estimated that he killed twenty thousand bison in a career as a hunter between 1870 and 1879 – twenty per day if he hunted three months of every year.[42] The Newton *Kansan* reported in 1872 that at least one thousand hide hunters roamed the southern plains.[43] If that number were in the grasslands for three months of the year, they could easily have supplied tanneries with over two million hides every year. Waste and the disruption of the bison's patterns of subsistence and reproduction could have raised their capacity for destruction higher still.

By what means could so many hides have been transported from the southern plains every year, if not on the Santa Fe, Kansas Pacific, and Union Pacific railroad lines that Dodge surveyed? The central plains railroads were the primary means of shipping bison hides to tanneries, but Dodge's exclusive focus on them

---

[35] Dodge, *Plains of North America*, 155.
[36] Strickland, "Recollections of W. S. Glenn." 25–31.
[37] Mayer and Roth, *Buffalo Harvest*, 49–52.
[38] Cook, *Border and the Buffalo*, 167.
[39] Wheeler, *Buffalo Days*, 80.
[40] Woody, interview with Haley, 11 February 1928, Panhandle-Plains Historical Society.
[41] Wayne Gard, *The Great Buffalo Hunt* (Lincoln: University of Nebraska Press, 1959), 128–129.
[42] J. Wright Mooar, interview with J. Evetts Haley, 25 November 1927, Panhandle-Plains Historical Society.
[43] Newton *Kansan*, November 28, 1872.

obscured humbler means of transportation. Many hide hunting outfits hauled their hides by the cartload not to railway stations, where dealers paid lower prices, but to markets in Texas, Colorado, and New Mexico. The hunter James Cator, for instance, hauled his hides by wagon to dealers in Las Vegas, New Mexico, and Granada, Colorado.[44] Dodge's estimate of the railroad's hide shipment was but a fraction of the number that were collected at temporary hide markets in the southern plains, such as Buffalo Gap in Taylor County and Hide Town in Scurry County, Texas. Like many mining towns in the Mountain West, these places flourished only briefly before being abandoned when the local resource became scarce.

By the mid-1870s, the army of hunters in the southern plains had pressed the herds south into the Texas Panhandle, far from the Kansas Pacific, Union Pacific, and Santa Fe lines. (See Figure 5.2.) The carnage in Texas in the mid-1870s was at least as extensive as that in Kansas in the early 1870s. Merry recalled that in the Texas Panhandle in 1876, "you could hear guns popping all over the country." To acquire the Panhandle hides, Charlie Rath, who had built a general store in Dodge City in July, 1872, and A. C. Reynolds founded a short-lived hide town alternately known as Rath City and Reynolds City. The town, a makeshift collection of buildings constructed of hides and sod, was briefly a center of the southern plains hide trade. Dealers bought over one hundred thousand hides at Rath/Reynolds City in the winter of 1876–77. "It was really a hide town," recalled Merry. "There were acres of ground covered with hides."[45] Another center of trade in the Panhandle was Fort Griffin, Texas, established as an Army post in 1867. John Irwin, a freighter for the government, compared the hide yards at Fort Griffin to "a great lumber yard. The hides were stacked tier upon tier in rows."[46] Far from the Kansas railroads, the hides that Euroamericans accumulated in the Texas Panhandle found their way out of the southern plains by the cartload.

The hide hunters' slaughter could not be sustained by a southern plains herd that numbered, at most, 15 million in the late 1860s. A herd of that size probably produced no more than four million calves a year. Wolves, accidents, drought, and competition from other grazers may have killed half that number in their first year. In 1871–72, a severe winter followed by drought destroyed, by one estimate, two hundred thousand head of cattle in western Texas. Although the loss of cattle reduced the competition for the bison's range, the blizzard and drought must have killed large numbers of bison as well. In October, 1873, a grass fire spread for 100 miles along the north side of the Kansas Pacific railroad, destroying millions of acres of forage.[47]

[44] Cator Family Papers, Panhandle-Plains Historical Society.
[45] Merry, interview with Haley, 21 August 1926, Panhandle-Plains Historical Society.
[46] John Chadbourne Irwin Papers, Southwest Collection, Texas Tech University.
[47] "Snow on the Plains," *New York Times* (13 December 1871), 2; "Great Loss of Cattle in Texas, Ibid. (29 March 1872), 1; "Incidents in Frontier Experiences – How the Monarch of the Plains is Hunted," Ibid. (2 April 1874), 3.

Figure 5.2. Hide hunters' camp in the Texas Panhandle, c. 1874. Courtesy of the Western History Collections, University of Oklahoma Library.

A bovine disease, Texas fever, may also have contributed to the demise of the bison in the southern plains. The disease, which affects grazing animals in warm climates, is caused by parasites transmitted by certain ticks belonging to the *Boophilus* genus. The parasites enter a host animal's bloodstream and attack the red blood cells, causing high fever, anemia, jaundice, and weakness. Texas longhorns were immune to the disease caused by the ticks they carried, but when they were driven to market they spread the affliction to Midwestern livestock. Like anthrax, which may have spread among Canadian bison in the 1820s and 1830s, Texas fever could spread without close contact between domestic and feral animals; the ticks could drop off their old hosts and attach themselves to new ones. Bison were susceptible to the disease. In 1905, when preservationists established a bison refuge in Oklahoma where longhorns had formerly grazed, they feared that the bison would be infected. Indeed, two of the first fifteen bison introduced to the refuge died of Texas fever.[48] The illness appeared as far north as Wyoming in 1885, brought by Texas cattle, but because the tick was susceptible to a hard frost, Texas fever was never well established in the northern plains. To the south, however, it may have contributed to the decline of the bison.

By the end of the 1870s, the combination of human and environmental pressures had reduced the southern plains bison to a few hundred stragglers. The destruction then moved north. Just as the extension of the Santa Fe Railroad to Dodge City spurred the expansion of commercial hide hunting in the southern plains, the extension of the Northern Pacific Railroad to Miles City, Montana Territory, in 1881, made possible the slaughter in the northern plains. "The prairies ... are covered with the carcasses of bison," wrote a *New York Times* correspondent from Miles City in April, 1880. Yet the hide hunters in the northern plains did not equal the destructiveness of their colleagues to the south. The *Times* correspondent estimated that the hunters brought in only ten thousand hides in the winter of 1880.[49] The naturalist William Hornaday thought that the slaughter was greater. He estimated that between 1881 and 1883, hide hunters in the northern plains shipped nearly three hundred thousand hides to the east.[50] The hide hunter Vic Smith wrote in his memoirs that the Montana hunters killed four hundred thousand bison in the winter of 1881–82.[51] Yet in 1884, the Northern Pacific carried only three hundred hides. Although hunters accumulated fewer hides than to the south, something certainly destroyed the northern plains bison. A rancher who traveled a thousand miles across the northern plains

---

[48] T. S. Palmer, "Our National Herds of Buffalo," *Tenth Annual Report of the American Bison Society, 1915–1916* (New York: American Bison Society, 1916), 44–46.
[49] "Montana's Indian Puzzle," *New York Times* (4 April 1880), 2.
[50] William T. Hornaday, *The Extermination of the American Bison, with a Sketch of its Discovery and Life History* (Washington, D.C.: Government Printing Office, 1887).
[51] Victor Grant Smith, *The Champion Buffalo Hunter*, ed. Jeanette Prodgers (Helena, Mt.: Twodot, 1997), 98.

in the early 1880s was, by his own account, "never out of sight of a dead buffalo and never in sight of a live one."[52]

In the northern plains, environmental pressures – both inherent in the grassland and anthropogenic – conspired with hunters to reduce the number of bison to a few stragglers. The decline of the northern plains bison resulted primarily from the combination of drought and ranches. The dominant shortgrasses of the western plains are well adapted to the drought-prone climate. By concentrating roots at the top of the soil, and keeping leafage minimal, shortgrass species conserve their moisture and take advantage of scarce rainfall.[53] Nonetheless, acute drought can devastate the shortgrasses upon which grazing animals such as the bison depend. When drought shrivels shortgrasses in the western plains, it also opens niches for them in the ordinarily more humid grasslands to the east. The mixed-grass plains – so-called because they lie between the semi-arid, shortgrass, western plains and the humid, tall grass, eastern prairies – are far enough from the Rocky Mountains that their dominant species, such as little bluestem, can grow to a height of one to two feet. During severe drought, however, little bluestem is reduced to a tiny proportion of the mixed-grass community, while shortgrass species, which thrive in relatively dry conditions, invade and dominate.[54] Thus in the historic plains drought did not so much destroy the bison's range as it temporarily shifted it to the east.

So long as bison were able to move eastward in search of forage, they could mitigate the impact of drought. By the late 1860s, however, they faced an obstacle. Lone Dog's winter count recorded the arrival of Texas cattle to the region north of the Platte River in 1868–69.[55] In succeeding years, bison increasingly faced competition for rangeland from domestic livestock. Between 1874 and 1880, the number of cattle in Wyoming increased from ninety thousand to over five hundred thousand.[56] By 1883, over five hundred thousand cattle grazed in eastern Montana.[57] Between 1874 and 1890, the cattle population in southern

---

[52] Theodore Roosevelt, *Hunting Trips of a Ranchman* (New York: G. P. Putnam's Sons, 1885).

[53] James K. Detling, "Processes Controlling Blue Grama Production in the Shortgrass Prairie," in *Perspectives in Grassland Ecology*, ed. Norman R. French (New York: Springer-Verlag, 1979), 25–39; Victor Shelford, *Ecology of North America* (Urbana: University of Illinois Press, 1963), 340; Philip L. Sims, J. S. Singh, and W. K. Lauenroth, "The Structure and Function of Ten Western North American Grasslands I: Abiotic and Vegetational Characteristics," *Journal of Ecology*, 66 (March 1978), 270; Sims and Singh, "The Structure and Function of Ten Western North American Grasslands II: Intra-Seasonal Dynamics in Primary Producer Compartments," *Journal of Ecology*, 66 (July 1978), 565; O. E. Sala and W. K. Lauenroth, "Small Rainfall Events: An Ecological Role in Semiarid Regions," *Oecologia*, 53 (June 1982), 301–304.

[54] Robert T. Coupland, "Effects of Changes in Weather upon Grasslands in the Northern Plains," in Howard B. Sprague, ed., *Grasslands* (Washington, D.C.: American Association for the Advancement of Science, 1959), 293.

[55] Mallery, "Picture-Writing of the American Indians," *Tenth Annual Report of the Bureau of American Ethnology* (Washington, D.C.: Government Printing Office, 1889) 285–286.

[56] Ernest Staples Osgood, *The Day of the Cattleman* (Chicago: University of Chicago Press, 1970), 53.

[57] William Cronon, *Nature's Metropolis: Chicago and the Great West* (New York: Norton, 1991), 220.

Alberta rose to one hundred thousand.[58] Domestic cattle not only encroached on the bison's range but occupied much of the mixed-grass region to which bison migrated during periods of drought. Drought struck the northern plains in the 1870s and early 1880s. Climatologists recorded droughts in eastern Wyoming and northeastern Colorado in 1873, 1879, and 1882. The handful of stations in Montana recorded low summer precipitation in 1877–78 and 1881–83.[59] During his two-year sojourn in the plains beginning in 1882, Theodore Roosevelt killed only a few bison for trophies, but his ranch on the North Dakota-Montana border was one of many that likely prevented northern plains bison from migrating to ranges in the mixed-grass plains to escape drought. Moreover, by occupying the bison's range, cattle made a recovery of the bison population unlikely.

Other inherent environmental causes of bison mortality – blizzard and grass fire – also took their toll on the northern plains bison in the 1870s and early 1880s. When Romanzo Bunn settled in Dakota Territory in 1880, he discovered the bleached skeletons of hundreds of bison in his half-section. The skeletons were largely intact, and Bunn found no broken bones or bullet-marked ribs among the remains. Bunn attributed the deaths to blizzard, such as the one he endured in his first winter in the northern plains in 1880–81. The blizzard that winter covered the plains with a foot of snow; drifts were 20 feet high in some places.[60] Grass fire in Alberta in 1878 destroyed many bison and drove survivors south into the Missouri River valley toward drought, the bullets of hide hunters, and the competition of domestic livestock.

Despite the influx of domestic cattle to the northern plains, exotic bovine diseases probably played less of a role in the decline of the bison there than they did to the south. A retired pathologist, Rudolph Koucky, suggested in 1983 that bovine disease was a factor in the destruction of the bison in the northern plains.[61] However, recent evidence suggests that brucellosis, a disease that causes spontaneous abortion, has only a minor impact on the bison's reproduction. Bison who survived the nineteenth century had high rates of infection of brucellosis, but because of the close contact usually necessary to transmit the disease, transmission more likely occurred not in the 1870s or 1880s but later, when ranchers corralled surviving bison in close proximity to domestic cattle.[62] Similarly, bovine tuberculosis, which was first diagnosed in bison at the Buffalo National Park in Wainwright, Alberta, in 1923, probably did not infect feral

---

[58] David. H. Breen, *The Canadian Prairie West and the Ranching Frontier, 1874–1924* (Toronto: University of Toronto Press, 1983), 65–66.

[59] Cary J. Mock, "Drought and Precipitation Fluctuations in the Great Plains During the Late Nineteenth Century," *Great Plains Research*, 1 (February 1991), 43–47.

[60] Bunn, "The Tragedy of the Plains," *Forest and Stream*, 63 (29 October 1904), 360–361.

[61] Koucky, "Buffalo Disaster of 1882," 23–30.

[62] Mary Meagher and Margaret E. Meyer, "On the Origin of Brucellosis in Bison of Yellowstone National Park: A Review," *Conservation Biology*, 8 (September 1994), 645–653; Meagher and Meyer, "Brucellosis in Free-Ranging Bison (*Bison bison*) in Yellowstone, Grand Teton, and Wood Buffalo National Parks: A Review," *Journal of Wildlife Diseases*, 31 (1995), 579–598.

bison in the Great Plains in the nineteenth century. Many of the bison at Wainwright descended from several hundred raised in Montana and purchased by the Canadian government in 1906. Most of the Montana herd went to Wainwright, but some went to Elk Island National Park. The Elk Island bison never developed tuberculosis, suggesting that the disease occurred only after the bison arrived in Wainwright. The infection probably came to Wainwright with a shipment of bison from a Canadian rancher who had raised captured bison calves by having his domestic cows nurse them – a common means of transmission of bovine tuberculosis.[63]

In the western plains, anthropogenic and environmental causes of bison mortality worked in concert. When Euroamerican hunters and ranchers added their pressures on the bison to the inherent causes of bison mortality in the dynamic plains, the bison population collapsed almost completely in only a decade and a half. When Hornaday surveyed the bison population in January, 1889, he discovered only twenty-five in the Texas Panhandle, twenty in the foothills of Colorado, ten between the Yellowstone and Missouri rivers, twenty-six near the Bighorn Mountains, and two hundred in Yellowstone National Park.[64] In just a few years, two dynamic forces – the plains environment and the American industrial economy – had combined to nearly obliterate the millions of bison that had inhabited the grasslands.

II

Economic behavior is embedded in culture, an assembly of principles and expectations that, while ever changing and often contested, set standards for conduct in the economic arena. As a consequence of the acceleration of industrialization in the second half of the nineteenth century, economic standards became the subject of political debate. In this period, Euroamericans argued over the legitimacy of labor unions, the promotion of railroad construction, the protection of industries from international competition, and the regulation of the money supply. As the scope of the hide hunters' destruction of the bison became clear, Euroamericans clashed over another issue: the wisdom of permitting the slaughter to continue. Those who opposed the destruction drew on two reform efforts of the post-Civil War United States: animal protection and Indian humanitarianism. The advocates of the slaughter appealed to an emerging belief that the extinction of the bison and the subjugation of Indians, however brutal, was necessary to open the Great Plains to Euroamerican settlement. The triumph of this group's beliefs permitted the rapid destruction of the bison in the 1870s and early 1880s.

[63] Seymour Hadwen, "Tuberculosis in the Buffalo," *Journal of the American Veterinary Medical Association*, 100 (January 1942), 19–22.
[64] Hornaday, *Extermination of the Bison*, 513.

The Destruction of the Bison

An enthusiasm for animal protection was the most important obstacle to the destruction of the bison. American efforts to prevent cruelty to animals began in the 1820s. Following the example of the Royal Society for the Prevention of Cruelty to Animals, founded in England in 1824, eighteen states enacted laws regulating the treatment of horses, sheep, and cattle between 1828 and 1861. Prominent antebellum reformers endorsed kindness to animals; Charles Lowell preached a sermon on the topic in Boston in 1837. Yet animal protection became a bona fide reform movement in the United States only after the Civil War. Henry Bergh, the son of a wealthy immigrant shipbuilder, founded the American Society for the Prevention of Cruelty to Animals in New York in 1866. Pennsylvania and Massachusetts SPCAs followed in 1867 and 1868, respectively, organized in both cases by wealthy, reform-minded women. By 1874, SPCA chapters were located in more than thirty of the largest cities in the Northeast and Upper Midwest.

The concentration of animal protection societies in the centers of American industry was not coincidental. Nineteenth-century animal protection was largely a reaction to industrialization. Nearly all reformers were middle-class city-dwellers. Critical and fearful of urban mechanization and poverty, they rued the loss of rural innocence, particularly farmers' solicitude for their animals. Preventing cruel treatment of animals in urban America would, they hoped, counteract the aggressive, competitive amorality of the marketplace. According to the historian James Turner, the advocates of animal protection "were conformists, comfortable, at heart happy with their up-to-date industrial world." Sensitive to the ill effects of industrialism, however, they "wanted to protest the sordid acquisitiveness of Victorian capitalism, but not too loudly."[65]

Animal protection was a decidedly feminizing movement, in its sizable female constituency and its invocation of a domestic ethic of anticruelty. Caroline Earle White in Philadelphia and Emily Appleton in Boston were early organizers of SPCAs in their cities, although both were formally excluded from the main sectors of the societies they helped to found. In 1869, White became the president of the Women's Branch of the Pennsylvania SPCA. In a short time, the Women's Branch surpassed the effectiveness, energy, and fund-raising ability of the original Pennsylvania SPCA.[66] Anticruelty was, above all, an explicit effort to extend the nineteenth-century ideology of feminine compassion to men and boys who were presumed to be responsible for most of the cruelties practiced upon animals. Pet-keeping and stories of the cruel treatment of innocent animals were intended to instill in boys an aversion to mistreating animals. Reformers were particularly concerned about boyhood hunting, which, they feared, established a taste for killing. Reformers thus portrayed hunting as a violent intrusion

---

[65] James Turner, *Reckoning with the Beast: Animals, Pain, and Humanity in the Victorian Mind* (Baltimore: Johns Hopkins University Press, 1980), 31–57.
[66] Ibid.

into the animal family. Harriet Beecher Stowe published a story in an American Sunday School Union tract that told of birds rendered motherless by the casual thoughtlessness of young hunters. The analogy to the human domestic sphere was obvious: to visit suffering on a family of animals was akin to violating the sanctity of the human family.[67]

Although anticruelty tracts criticized hunting, they focused their attentions largely on urban and suburban animals: pets, carriage horses, and stray dogs. To prevent the hunting of wild animals was, at first, not a significant part of their mission. But the destruction of the bison excited Bergh to action. In the early 1870s, he received a number of letters decrying the slaughter. "We speak on behalf of the buffalo, antelope, and various wild game of the western prairies," a group of women from Freeport, Illinois, recently returned from a trip through the plains, wrote to Bergh in December, 1872. "Visiting these sickening scenes of slaughter, we find the Plains thickly strewn with carcasses of buffalo, deer, and antelope."[68] High-ranking army officers stationed in the plains also wrote to Bergh objecting to the slaughter. When General William Hazen and Lieutenant-Colonel A. G. Brackett protested to Bergh, he forwarded copies of the letters to ASPCA members, newspapers, and Congressional supporters, hoping to spur the creation of legislation to regulate the hide hunters.

The officers' letters in particular revealed how the effort on behalf of the bison struggled to extend the rhetoric of anticruelty – previously applied only to farm animals, pets, songbirds, and some urban draft animals – to a wild animal. Hazen described the bison as "a noble and harmless animal, timid, and as easily taken as a cow." He called the hunt a "wicked and wanton waste." Brackett also took care to present the bison as both "harmless and defenseless": a gentle, nearly domesticated animal. (See Figure 5.3) "[T]here is as much honor and danger in killing a Texas steer as there is in killing a buffalo," wrote Brackett. "It would be equally as good sport," he argued, "to ride into a herd of tame cattle and commence shooting indiscriminately." The officers' analogies to domestic animals, which were protected in many states by anticruelty legislation, and their conclusion that the killing of bison was not simply needless but "wicked" (according to Hazen) and, more to the point, "cruel" (according to Brackett), revealed their debt to the arguments of the anticruelty advocates.[69] The officers' letters also

[67] Katherine C. Grier, "'Kindness to All Around': A Domestic Ethic of Kindness to Animals, 1820–1870," Paper presented to the Shelby Cullom Davis Center for Historical Studies seminar "Animals and Human Society," Princeton University, Princeton, N.J., April 25, 1997. Cited by permission of the author.
[68] Edward P. Buffett manuscript, "Bergh's War on Vested Cruelty," vol. 6, American Society for the Prevention of Cruelty to Animals Archives, New York. See also Zulma Steele, *Angel in Top Hat* (New York: Harper, 1942), 162.
[69] "Slaughter of Buffaloes," *Harper's Weekly*, 16 (24 February 1872), 165–166. A. G. Brackett, "Buffalo Slaughter," *New York Times* (7 February 1872). "Protection of Buffalo" *Congressional Record* (March 10, 1874), 2106.

THE LAST BUFFALO.

"Don't shoot, my good fellow! Here, take my 'robe,' save your ammunition, and let me go in peace."

Figure 5.3. Thomas Nast, "The Last Buffalo," Harper's Weekly, June 6, 1874

revealed that the extension of SPCA rhetoric of domestic kindness to a wild animal was expanding the organization's ideology in unanticipated ways.

Animal protection was a relatively new passion in the 1870s. So, too, was Indian humanitarianism. The 1870s were liminal years in United States Indian policy, marking a transformation from a policy of separation to one of assimilation. From the beginning of the nineteenth century until the Civil War, policymakers sought to confine Indian populations to reservations in order to segregate them from Euroamerican settlers. Indian policy in the 1870s was, in one sense, a continuation of this program. "Our civilization is ever aggressive, while the savage nature is tenacious of traditional customs and rights," wrote Columbus Delano, the Secretary of the Interior, in 1873. "This condition of things calls loudly for more efficient efforts to separate Indians from whites by placing them on suitable reservations."[70] Yet in the years after the Civil War, the United States also embarked on a so-called "peace policy," a cooperative effort between the federal government and Protestant missionaries to Christianize and educate reservation Indians. The peace policy foreshadowed the ambitious, albeit deeply flawed program of Indian assimilation that began in the 1880s and continued into the twentieth century.[71]

To a considerable extent, Indian humanitarianism emerged from the United States' experience in the Civil War. Tired of bloodshed, the editors of the *New York Times* condemned an army attack on a Piegan winter camp in 1870 as a "sickening slaughter" – the same terms the Freeport women used to describe the killing of bison. To subdue Indians by violence imperiled "our standing before the world as a Christian nation." The heightened interest in racial justice for recently emancipated slaves also influenced Indian humanitarianism. "We have long been doing justice to the negro," the *Times* wrote later in 1870, on the occasion of a visit to Washington by the Oglala leader Red Cloud. "Is it not almost time to see what we can do for the Indian?"[72]

The humanitarianism of the peace policy was not, however, a complete departure from the history of hostilities between Indians and Euroamericans. "What we must do if we mean to save the remnant of the Indians," the *Times* editorialized in 1870, "is to gather them all into a small district which we can really police and protect, and there teach them the arts of civilized life." What the *Times* euphemistically termed "gathering" was, in plain words, the application of military force. Indeed, the editors endorsed military efforts to confine the plains nomads to reservations; they approved of the 1876 campaign to subdue the "hostile," off-reservation Sioux led by Crazy Horse and Sitting Bull. As far as the reformers were concerned, only reservation Indians deserved the benefits of material support and Christian education. Those nomads who continued to resist the pressure to go to reservations, the *Times* concluded, remained "utterly out of the reach of every humanizing influence." Following the Sioux defeat of Custer at the Little Bighorn in June, 1876, this already straitened channel of humanity was closed. In the wake of Custer's death, the *Times* counseled that the off-reservation Sioux deserved "condign" punishment.[73]

---

[70] *Annual Report of the Secretary of the Interior, 1873*, 43rd Cong., 1st Sess. (Serial 1601), ix.

[71] For nineteenth-century Indian assimilation, see Frederick E. Hoxie, *A Final Promise: The Campaign to Assimilate the Indians, 1880–1920* (Lincoln: University of Nebraska Press, 1984); Henry E. Fritz, *The Movement for Indian Assimilation, 1860–1890* (Philadelphia: University of Pennsylvania Press, 1963).

[72] "The Slaughter of the Piegans," *New York Times* (24 February 1870), 4; "The Piegan Slaughter and Its Apologists," Ibid. (10 March 1870), 4; "The Montana Massacre-Col. Baker's Report," Ibid. (12 March 1870), 4; "The Last Appeal of Red Cloud," Ibid. (17 June 1870), 4.

[73] "The Oldest of American Difficulties," Ibid. (22 May 1870), 4; "The Indian War," Ibid. (17 July 1876), 2.

Although all humanitarians wanted Indians of the western plains eventually to renounce nomadic bison hunting in favor of agriculture, some strongly advocated immediate cessation of the slaughter of the bison. They believed that such destruction was counterproductive, driving desperate Indians to acts of violence. The editors of *Harper's Weekly* contended in 1874 that "the indiscriminate slaughter of the buffalo … has been the direct occasion of many Indian wars. Deprived of one of their chief means of subsistence through the agency of white men, the tribes naturally take revenge by making raids on white settlements and carrying off stocks."[74] *Harper's* asked its readers to take an unprecedented step: to extend the sentimentality and morality of animal protection and Indian humanitarianism from peaceful reservation Indians to off-reservation hunters, and from domestic to wild animals. More typical was the view of the *Nation*, however, which like *Harper's* attributed Indian violence to the demise of the bison. Far from criticizing the destruction of the nomads' resources, however, the *Nation* urged that nonreservation Indians first be "hunted down," and then subjected to a redoubled effort at cultural assimilation.[75] The wild Indians of the plains required taming.

These contradictions in the reformers' approach to Indians, which would eventually be the undoing of the effort to halt the hide hunters, were embodied in the bison's unlikely legislative champion, United States Representative Greenburg Lafayette Fort. A Republican from central Illinois, Fort's four terms in the House between 1873 and 1881 were the pinnacle of his political career. Two characteristics distinguished Fort's tenure in Congress. First, he was a practitioner of symbolic politics: the introduction of divisive, openly partisan legislation that stirred his supporters, irritated his opponents, and had little hope of passage. Fort, like many Republicans, most often engaged in such political symbolism when he "waved the bloody shirt," a postwar euphemism for the appeal to wartime loyalties. A Union veteran who had answered the call for volunteers in April, 1861, Fort introduced a resolution requiring members of the House to give preference to wounded Union soldiers when appointing aides and subordinates. He repeatedly proposed legislation granting pensions to all disabled, honorably discharged Union veterans. Although his bills ultimately failed, they galvanized Fort's supporters and won him national recognition from Union veterans.

Second, Fort saw himself as a reformer, while many of his colleagues saw him as no more than a crank. His oftentimes inane motions, amendments, and objections ("I object to everything that is out of order," he once said in debate) provoked the exasperation of the chamber. Arriving in Washington to take his seat in Congress in 1873, when Republican strength and enthusiasm for Reconstruction

---

[74] *Harper's Weekly* (12 December 1874), 1022–1023.
[75] "Our Indian Wards," *The Nation* (13 July 1876), 20–22.

in the South was waning, he initially turned his attention to the reform of the Western territories. The West in the 1870s was riddled with corruption. Territorial governments and the Indian Office were filled with inept political appointees; politicians at all levels of government enriched themselves by corrupt dealings in railroad construction. Indeed, Western railroad graft reached the highest levels of the Grant administration. Fort introduced bills – none of which became law – designed to prevent territories or municipalities in the territories from entering into debt to finance railroad construction; to create a new government for the Indian Territory; and to eliminate the corruption of territorial governments entirely by rapidly admitting territories to statehood. By 1877, however, Fort had shifted his reformist energies to the advocacy of paper currency and the coinage of silver. Although these positions put him at odds with the tight money policies of leading Republicans, they were politically astute stances for a representative from rural Illinois where farmers organized powerful political lobbies opposing the deflation of the currency.[76] Fort's ultimate shift from Western issues to inflation demonstrated that his reformism, although probably genuine, was nonetheless flexible enough to appeal to the shifting concerns of his constituents.

At almost every turn, Fort's grandstanding encountered the opposition of Columbus Delano, the Secretary of the Interior between 1870 and 1875. Fort regularly seized on popular causes to further his appeal, but Delano's career was an embattled one, characterized by his unflinching advocacy of unpopular positions. Before joining Grant's cabinet, Delano served three terms as an Ohio Congressional representative; from 1845–47 as a Whig from a largely Democratic district, and from 1865–69 as a Republican. Like Fort, he shaped his political views around sectional issues, antagonizing not only Southerners but Northern Democrats in the process. Unlike Fort, he was not adept at bolstering his popularity among his constituents. In his first term in Congress, he opposed the Mexican War as a Southern conspiracy. After the Civil War, he often sided with the radical faction of House Republicans on issues concerning the Reconstruction of the South. After leaving the House to join the Grant administration, as Commissioner of Internal Revenue from 1869 to 1870 and thereafter as Secretary of the Interior, Delano brought contention with him. Internal Revenue and Interior were two of the most venal branches of the federal government in a period distinguished for its extensive public corruption. Although Delano was probably an honest man – a rare commodity in Grant's cabinet – he was unmindful of his subordinates' corruption and uninterested in rooting it out. The liberal press and backbenchers such as Fort therefore subjected Delano to a barrage of criticism for his failure to weed out corruption. Criticism of Delano peaked in 1875. In April, the *New York Tribune* accused the Secretary's son,

---

[76] *Congressional Record*, 44th Cong., 1st Sess. (January 31, March 13, April 24, 1876), 773, 1678, 1813; Ibid., 45th Cong., 1st Sess. (November 23, 1877), 622–631.

John, of benefiting from corrupt Interior Department transactions in Wyoming. That summer, the House investigated Delano's supervision of contracts to supply Indian reservations with beef. Although concluding that Delano had awarded contracts to high bidders, the House exonerated him of any technical wrongdoing.[77] In September, however, the Yale paleontologist O. C. Marsh renewed the charges against Delano during an interview with Grant. The *Nation* reported on the exchange and called for Delano to resign. Under unrelenting pressure, Delano offered his resignation, but Grant refused to accept it. Delano therefore soldiered on in a job he no longer desired, inured to criticism, resigned to his unpopularity, and, quite unlike Fort, unmovable in his policies despite his many detractors in the government and the press.[78]

The slaughter of the bison brought Fort and Delano into direct opposition. Fort introduced bills in Congress in 1874 and 1876 making it "unlawful for any person who is not an Indian to kill, wound, or in any manner destroy any female buffalo, of any age, found at large within the boundaries of any of the Territories of the United States." Fort's idiosyncratic brand of reformism and symbolic politics, although alienating members of both parties, also garnered initial widespread, if shallow, support for his bison legislation. As symbolic politics, the bills offered Fort's reform-minded colleagues in the House the opportunity to castigate the Secretary of the Interior on another matter related to Indian subsistence. To Indian humanitarians, the bills restored the integrity of the treaties of Medicine Lodge and Fort Laramie. To anticruelty advocates, the bills outlawed an ignoble carnage. Reformers in both parties supported the bills. In 1874, when Republicans were in the majority in the House, and 1876, when Democrats were in the majority, the Committee on the Territories unanimously approved the bills and a majority of the House supported the legislation; a majority of the Senate voted in favor of the 1874 bill.[79]

The bills drew support because they marshaled the arguments for both animal protection and Indian humanitarianism. Fort made his sympathies plain in 1876, saying "For my part I favor the society which has in view the prevention of cruelty to animals." Most legislators who spoke in favor of the bills were Northeastern or Midwestern reformers with similar sentiments. George Hoskins, a Republican from New York, perhaps taking his cue from the letter of Lieutenant-Colonel Brackett, which was read into the record during the course of the 1874

---

[77] "Contracts for Indian Supplies and Transportation for the Fiscal Years Ending June 30, 1873, and June 30, 1874," House Report No. 778, 43rd Cong., 1st Sess. (Serial 1627).

[78] Daniel W. Delano, Jr., *Franklin Roosevelt and the Delano Influence* (Pittsburgh: James S. Nudi, 1946), 101–133. Allan Nevins, *Hamilton Fish: The Inside History of the Grant Administration*, vol. 2 (New York: Frederick Ungar, 1936).

[79] In 1874, House Republicans held a 100-seat majority, and Fort's bison bill passed by a voice vote. In 1876, House Democrats held a 60-seat majority, and the bison bill passed by a nearly two-to-one vote. For the debates on the bison bills, see *Congressional Record* (March 10, 1874), 2105–2109; "Slaughter of Buffaloes," Ibid. (February 23, 1876), 1237–1241.

debate, compared Fort's bill to extant state game- and animal-protection laws. Fort, who preferred a hyperbolic style of debate, described the killing of bison as "wanton wickedness," a term usually found in the SPCA's tracts. One of Fort's reform-minded colleagues in the House, the Republican Joseph Hawley of Connecticut, brought the issue closer to home, alluding both to motherhood and rural innocence in describing a hide hunter as a man likely to "shoot down his mother's cow in the barn-yard." Invoking the high moral ground of the SPCA, the supporters of Fort's bill portrayed the killing of bison as a thoroughly sordid business. Supporters of the bill further appealed to the arguments of Indian humanitarians. In his opening address on his 1874 bill, Fort articulated the strongest argument of the humanitarians: "I am not in favor of civilizing the Indian by starving him to death." Fort's colleague, the Republican David Lowe of Kansas, chimed in that the destruction of the nomads' means of subsistence "will not do in this age of civilization and Christianity."

On the floor of the House the question of animal protection was morally – or at least politically – unassailable, but Indian policy was a nettlesome problem. Fort's supporters were uncomfortable as partisans of the Indians. Lowe denied that his support of the bill was "simply a matter of sentiment in behalf of the Indian." In 1876, Fort declared that "I have no especial sentimentality in my bosom for the Indian; I have no especially friendly feeling for the savage." In the course of the 1874 debates, the Ohio Republican James Garfield – who six years later would be elected president – echoed Secretary Delano, posing the question that ultimately derailed Fort's legislation. However cruel the actions of the hide hunters may be, he pointed out, they were destroying the primary resource of the plains nomads and thereby forcing them to reservations. "If the barbarism of killing buffalo for mere wanton sport has any compensation in it," he suggested, "perhaps it may be this is a compensation worthy of our consideration."[80]

The effort to confine Indians of the plains to reservations posed a dilemma to Indian humanitarians. Forcing Indians to reservations and the aims of the peace policy were indeed complementary. Quite apart from the peace policy's hostility to Indian cultural traditions, it was not the entirely benign program that its appellation suggested. As Delano explained in 1873, the "so-called peace policy sought, first, to place the Indians upon reservations as rapidly as possible." Only on reservations were Indians treated to the government-sponsored advice and aid of missionaries. Toward nonreservation Indians, the federal government's policy was punitive. Delano wrote, "whenever it is found that any tribe or band of Indians persistently refuse to go upon a reservation and determine to continue their nomadic habits ... then the policy contemplates the treatment of such tribe or band with all needed severity ... thereby teaching them that it is better to follow the advice of the government." Delano saw the destruction of the bison as

---

[80] *Congressional Record* 43rd Cong., 1st Sess. (March 10, 1874), 2106–2107.

a means of pacification: "The rapid disappearance of game from the former hunting-grounds must operate largely in favor of our efforts to confine the Indians to smaller areas, and compel them to abandon their nomadic customs."[81]

Delano interpreted the treaties of Medicine Lodge and Fort Laramie as permitting the destruction of the bison by Euroamerican hide hunters. He applauded the results of their efforts, writing that he "would not seriously regret the total disappearance of the buffalo from our western prairies, in its effect on the Indians, regarding it rather as a means of hastening their sense of dependence upon the products of the soil and their own labors." He deplored the presence of hide hunters on Indian reservations but noted that the Treaties of Medicine Lodge did not exclude them from the area south of the Arkansas River designated as the hunting territory of the southern plains nomads. In regard to the northern plains, he regretted the stipulation of the Treaty of Fort Laramie of 1868 granting the Sioux the exclusive right to hunt north of the Platte River so long as the bison remained there in significant numbers. He therefore stated – quite unilaterally – that the number of bison ranging north of the Platte was insufficient to support the Sioux, and recommended in both his 1873 and 1874 reports that Congress abrogate that article of the treaty.[82] Congress did not act, so in 1875, Delano and Grant told a Sioux delegation to Washington led by Red Cloud and Spotted Tail that the region between the Platte and Niobrara rivers no longer contained enough bison to support the Sioux and must be ceded to the United States.[83]

Delano had many sympathizers in Congress. In 1874, the Missouri Republican Isaac Parker called his policy sound. In 1876, the Texas Democrat John Hancock explicitly rejected humanitarianism in favor of Delano's policies. "I hope, sir, there is no humanitarian sentimentality that would induce legislation for the protection of the buffalo," he said to Fort. "If the theory upon which the Government is now treating the Indians is a proper one, and I am inclined to believe it is the best, the sooner we get rid of the buffalo entirely the better it will be for the Indian and for the white man, too." Although Fort's support was strong enough to ensure passage in both houses in 1874 and in the House in 1876, well-placed opponents defeated the bills. President Grant, who had made his support for Delano's policy plain to Indian delegations to Washington, killed the 1874 bill with a pocket veto.[84] In February, 1876, Fort's bill passed in the House but never came to a vote in the Senate Committee on the Territories. After Custer's defeat at the Little Bighorn in late June, the bill had no chance of passage.

---

[81] Delano, *Annual Report of the Secretary of the Interior, 1873*, iii–ix; Ibid., *1872*, 42nd Cong., 1st Sess. (Serial 1560), 5.
[82] Ibid., *1873*, vi–viii; Ibid., *1874*, 43rd Cong., 2nd Sess. (Serial 1639), xi.
[83] "The President's Last Interview with the Sioux," *New York Times* (3 June 1875), 1.
[84] "Slaughter of Buffaloes," *Congressional Record*, 44th Cong., 1st Sess. (February 23, 1876), 1239; "Red Cloud: The Indians in Washington Visit the White House," Ibid. (29 May 1872), 1; "The Sioux and the White House," Ibid. (27 May 1875), 5.

Delano's policy of relying on the exhaustion of the herds in order to relegate Indians to reservations was entirely consistent with a mid-century image of the bison that was radically different from the SPCA's depiction of a gentle ruminant. Popular accounts of the bison described a violent species consistently losing its battles against human hunters and the forces of nature. This view of the bison pervaded an 1869 description in a popular journal. Theodore Davis, the author of the article, argued that "the buffalo is fast disappearing." To the pressure of Euroamerican hunters Davis added Indian hunters and environmental factors: blizzard, drowning, and the predation of wolves. The bison competed among themselves as well. Describing the contests between bulls to establish a rank order during the rutting season, Davis wrote: "The bulls fight viciously, and are attended during these combats by an admiring concourse of wolves, who are ever ready to come in at the death of either of the combatants, or will even take a chance in and finish any killing that has been imperfectly done." One of the illustrations accompanying the article showed a "Battle for Life": a bison fighting off an attack by a pack of wolves.[85] In the 1860s and 1870s, such scenes were a staple of popular depictions of the bison.[86] (See Figure 5.4.)

The inevitability of the bison's extinction pervaded an 1876 study of the species by Joel Allen, a Harvard University zoologist. Allen's book was a hybrid of neo-Lamarckian and natural selection theories and a combination of scientific and popular accounts of the bison. Allen came to his subject while analyzing the fossil remains of extinct species of bison; the extirpation of *Bison latifrons* and *Bison antiquus* foreshadowed the fate of *Bison americanus*. His study was largely a recitation of the bison's disappearance from one region of North America after another, owing to both Euroamerican and Indian hunters and environmental pressures. He called the species "sluggish" – in other words, unfit – and concluded that "the period of extinction will soon be reached."[87] Allen's views exemplified both the popular and academic understanding of Charles Darwin's theories of natural selection. Most Americans conflated Darwin's ideas with those of Herbert Spencer and the neo-Lamarckian naturalists such as Edward Cope and Alpheus Hyatt, viewing natural selection as a process that weeded out the unfit in favor of ever-more-excellent forms of life. According to the neo-Lamarckians, the process leading from bison to cattle paralleled the replacement of Indians by Euroamericans.[88] Ten years after the hide hunters' slaughter ended,

---

[85] Theodore R. Davis, "The Buffalo Range," *Harper's New Monthly Magazine*, 38 (January 1869), 147–163. The predation of wolves was real, but late twentieth-century zoologists and wildlife biologists attest that contests between bulls to establish rank order are brief and almost never deadly.

[86] See Larry Barsness, *The Bison in Art: A Graphic Chronicle of the American Bison* (Fort Worth, Tex.: Amon Carter Museum of Western Art, 1977), 62–63, 80–81.

[87] Allen, *American Bisons*, 35, 56, 180–181.

[88] Peter J. Bowler, *Evolution: The History of an Idea*, rev. ed. (Berkeley: University of California Press, 1989), 24, 258. For the neo-Lamarckians, see also Edward J. Pfeifer, "United States," in Thomas F. Glick, ed., *The Comparative Reception of Darwinism* (Chicago: University of Chicago Press, 1988), 168–206.

Figure 5.4. William M. Cary, "Buffalo Bulls Protecting a Herd from Wolves," Harper's Weekly, August 5, 1871.

the University of Wisconsin historian Frederick Jackson Turner formalized this Spencerian interpretation of western settlement in his influential essay, "The Significance of the Frontier in American History."[89] Turner's interpretations of the history of the American West prevailed well into the twentieth century.

Lawmakers friendly to Delano's Indian policy embraced the idea that the destruction of the bison was an inevitable part of the advance of civilization. They were particularly attracted to the notion that domestic cattle made a higher use of the range than bison. Representative Omar Conger of Michigan said in 1874 that "there is no law which a Congress of men can enact, that will stay the disappearance of these wild animals before civilization. They eat the grass. They trample upon the plains upon which our settlers desire to herd their cattle and their sheep.... They are as uncivilized as the Indian."[90] Hancock of Texas seconded that sentiment in 1876; saying the bison "are, at most, but game. Men have not been able to domesticate them so as to make them useful in any respect as a domestic animal. They take up as much room and consume as much provender as cattle and horses or any other character of useful domestic animals."[91] Relying on popular and scientific accounts of the bison's tenacity to contrast the bison with familiar domesticated species, Conger and Hancock rejected the SPCA's depiction of the bison as nearly tame. They argued, in short, that the untamed bison was the resource of the savages; the bison's displacement by domestic cattle was a victory for civilization.

Ratification of Fort's bills may not have mattered, because most army officers in the plains were disinclined to enforce them. They were as enthusiastic as Delano about the disappearance of the bison. Colonel Richard Irving Dodge did not prevent hide hunters from entering the Indians' preserve south of the Arkansas River. In 1873, he reportedly told Josiah Wright Mooar, who had come to inquire about the propriety of hunting south of the Arkansas, "If I were a buffalo hunter, I would hunt where the buffaloes are."[92] When Sir William F. Butler, a British officer, confessed to Dodge that he had shot more than thirty bison on the North Platte, Dodge was gleeful. Butler wrote, "I could not but feel some qualms of conscience at the thought of the destruction of so much animal life, but Colonel Dodge held different views. 'Kill every buffalo you can,' he said; 'every buffalo dead is an Indian gone.'"[93]

The army adhered to Delano's policy of allowing the exploitation of the herds to proceed unfettered. Even some hide hunters ascribed to this view of the slaughter. "The buffalo didn't fit in so well with the white man's encroaching

---

[89] Frederick Jackson Turner, "The Significance of the Frontier in American History," *American Historical Association Annual Report* (1893), 199–227.
[90] *Congressional Record*, (March 10, 1874), 2107.
[91] *Congressional Record*, (February 23, 1876), 1239.
[92] J. Wright Mooar, "Buffalo Days," as told to James Winford Hunt, *Holland's*, 52 (February 1933), 10, 44.
[93] *Sir William Butler: An Autobiography* (New York: Scribner, 1913), 97.

civilization," argued Mayer. "So he had to go."[94] Although he recognized that hide hunters deprived the plains nomads of their subsistence, John Cook reflected that the "ruthless slaughter" was "simply a case of the survival of the fittest. Too late to stop and moralize now."[95] Eastern reformers reluctantly agreed. In the Black Hills, the *New York Times* editorialized in 1875, "the red man will be driven out, and the white man will take possession. This is not justice, but it is destiny."[96] These rationalizations rested on the assumption of the innate superiority of Euroamericans and their land use strategies. The belief in Euroamerican progress assumed an inevitable advancement toward higher forms: from Indians and bison to Euroamericans and domestic livestock. For Delano, as for Sherman – who had been among the first to predict the demise of the bison – extinction of the bison was not only an inevitability but a solution to the problem of domesticating the western plains.

III

The political economy of the 1870s pitted not only Euroamerican bison hunters against the plains nomads, but Euroamericans against each other. Bison hunters rarely encroached upon each others' hunting territory, but they often found themselves engaged in usually losing battles against hide dealers, freighters, and retailers. Exorbitant prices for supplies and a glutted market in hides in the mid-1870s broke many bison hunters. The destruction of and competition for the bison not only deprived the plains nomads of their primary resource, but also deprived many hide hunters of the rewards of their labors.

Josiah Wright Mooar was among the first hunters to turn to the procurement of hides. Beginning in 1870, he provided bison meat at three cents a pound to construction workers on the Santa Fe and Kansas Pacific railroads and soldiers at Fort Dodge.[97] When the railroads laid off their construction workers for the season in the fall of 1872, many entered the field as bison hunters and skinners.[98] Theodore Raymond and the brothers Ed and Bat Masterson were typical; they came to the southern plains in 1872 with a grading subcontract for the Santa Fe Railroad before turning to bison hunting.[99] Richard Bussell, a "bull-whacker" hauling freight between Fort Laramie, Salt Lake City, and Leavenworth, Kansas,

---

[94] Mayer and Roth, *Buffalo Harvest*, 27–28.
[95] Cook, *Border and the Buffalo*, 166–167.
[96] "Taming the Savage," *New York Times* (15 April 1875), 6.
[97] Odie B. Faulk, *Dodge City: The Most Western Town of All* (New York: Oxford University Press, 1977), 31.
[98] William E. Connelley, ed., "Life and Adventures of George W. Brown, Soldier, Pioneer, Scout, Plainsman, and Buffalo Hunter," *Collections of the Kansas State Historical Society*, 17 (1926–28), 116–117.
[99] Raymond, "Diary of a Buffalo Hunter, 345–346.

switched from freighting to bison hunting in 1868. His forays into hunting were sporadic, however; he continued hauling intermittently until 1872. Like the other bison hunters in the southern plains, he gradually drifted south in pursuit of the herds, moving from Dodge City to Fort Griffin in the Texas Panhandle. Often unable to sell his hides for a suitable price in Fort Griffin, Bussell hauled them to Dallas. Even so, Bussell lamented that he earned less for his labors than he had hoped.[100] Bussell's experiences exemplified bison hunting in many ways: transitory, episodic, and often unprofitable. Yet to poor men, bison hunting seemed to promise great and instant wealth. By the winter of 1872–73, according to the Newton *Kansan*, between one and two thousand hide hunters were pursuing bison in western Kansas. The economic depression that began in 1873 drove still more hunters into the southern plains. The rancher Charles Goodnight estimated that there were three thousand hunters in the Texas Panhandle in the mid-1870s.[101]

For a few years, the slaughter produced tremendous profits, but not necessarily for the hunters. From 1870 to early 1872, the price of a bull hide stood at $3.50, for that of a cow and calf slightly less. Frank Mayer calculated that inasmuch as cartridges cost 25 cents each, "every time I fired one I got my investment back twelve times over." If he could kill one hundred bison every day – which did not seem an unreasonable figure to him when he embarked on his career as a bison hunter – he figured to earn $200 a day. Mayer vastly overestimated the number of bison he could kill in a day, however. He also failed to consider that commercial hide hunting required a considerable investment: $400-$650 for a good wagon and $125 for a .50-caliber Sharp or Remington rifle. Altogether, it cost Mayer $2,000 to outfit himself for the hunt.[102] The large initial investment meant that many independent hunters went broke. Others judged themselves fortunate just to break even.[103]

Even after outfitting themselves, hunters, like miners in gold-rush boomtowns, paid inflated prices for ordinary goods and services. Prices in Dodge City were scandalously inflated. Wentin Wilson reported in 1876 that in Dodge, "everything is high: 25 cents for a shave; 75 cents for a haircut, whiskey 25 and 50 cents a drink."[104] The builders of a wooden bridge across the Arkansas at Dodge City charged bison hunters $1.50 to haul their hides across the river and into town.[105]

---

[100] Bussell, "Hunting Buffalo in the Panhandle," Panhandle-Plains Historical Society.
[101] Claude *News*, 13 March 1931. Bussell estimated that between fifteen thousand and twenty thousand men were engaged in the business of bison hunting – a not entirely unreasonable figure if one includes not only shooters but skinners, haulers, dealers, and the various merchants in the hide towns. Bussell, "Hunting Buffalo in the Panhandle," Panhandle-Plains Historical Society.
[102] Mayer and Roth, *Buffalo Harvest*, 49–52.
[103] Connelley, ed., "Life of George W. Brown," 111, 123.
[104] Wilson, "Details of a Buffalo Hunt of 1876 in Kansas," Kansas State Historical Society, Topeka.
[105] Connelley, ed., "Life of George W. Brown," 118.

The economic insecurities of hunting outfits inspired one anonymous bison skinner in the 1870s to modify the lyrics to "Canada-I-O," a Northern New England folksong which laments the hardships of seasonal work logging in Quebec. The revised song, "Buffalo Range," tells of an unemployed Texan who is enticed by the promise of "good wages" to join a hunting outfit headed for the Panhandle. After the outfit endures Indians' bullets, mosquitoes, and poor rations, however, the merchant who had organized the outfit declares bankruptcy. "Now we're back across Peace [sic] River and homeward we are bound," the song concludes, "In that forsaken country may I never more be found. / If you see anyone bound out there pray warn them not to go / To that forsaken country, the land of buffalo."[106]

The lament of the skinner in "Buffalo Range" would have been familiar to James Cator. Although Cator did not keep an account book, he saved many of his receipts from his bison hunting career between 1874 and 1878. In his first year as a hunter, he almost certainly failed to break even. His receipts – probably incomplete – show that he sold nearly $1,100 worth of hides to A.C. Myers and Charles Rath of Dodge City, but he also purchased nearly $2,900 worth of supplies. In 1875 and 1876, he looked to increase his profits by selling not only hides but meat, tongues, and the occasional wolf or coyote pelt. As prices dropped, Cator, like Bussell, hauled his hides to competing, far-flung dealers in search of the best price, selling his hides in New Mexico and Colorado.[107] By the late 1870s, Cator had abandoned hunting and, with his brother, established a trading post on one of the main wagon trails connecting Dodge City to the Texas Panhandle.[108]

Bussell and Cator were fortunate in that they hauled their own hides to Dodge City, Fort Griffin, or wherever they could find the best price. Freighters and merchants often went directly to the outfits camped on the plains to eliminate competition.[109] The hide dealer Charles Rath dispatched teams to the bison range to buy hides directly from hunting outfits for as little as 75 cents per hide.[110] At Rath/Reynolds City in the Texas Panhandle, prices were extortionate; tobacco sold at $2 per pound, corn at 5 cents a pound. The dealers, meanwhile, paid from 50 to 90 cents for bison hides that sold for perhaps twice as much in Fort Griffin. The dealers nonetheless accumulated large numbers of hides. A visitor reported "buffalo hides piled up here as high as a house."[111]

As hunters flooded the market with hides, the value of the product steadily declined. A tanner who experimented with bison hides in 1871 paid Josiah Wright

---

[106] N. Howard Thorp, *Songs of the Cowboys* (New York: Clarkson N. Potter, 1966), 195–218.
[107] Cator Family Papers, Panhandle-Plains Historical Society.
[108] C. Robert Haywood, *Trails South: The Wagon-Road Economy in the Dodge City-Panhandle Region* (Norman: University of Oklahoma Press, 1986), 107–108.
[109] J. W. Woody, interview with Haley, 19 October 1926; John Wood interview with L. F. Sheffy, Canadian, Texas, 28 December 1929, Panhandle-Plains Historical Society.
[110] Bussell, "Hunting Buffalo in the Panhandle," Panhandle-Plains Historical Society.
[111] Wilson, "Details of a Buffalo Hunt," Kansas State Historical Society.

Mooar $2.25 per hide. In 1872, before the rush of hunters into the southern plains had begun in earnest, Mooar sold his hides to a tannery in Pennsylvania for $3.50 per hide.[112] By 1874 and continuing until 1876, however, the price of a bull hide in the glutted market had fallen to $1.15, and that of a cow or calf hide to 65 cents.[113] In December, 1874, Dodge City dealers paid James Cator $2.15 for his bull hides. Two years later, the price had fallen to 80 cents per hide.[114]

The falling value of the hides drove some hunters to greater devastation by expanding their carnage to include the poisoning of wolves, foxes, and coyotes. The hunters stocked up on poison – usually strychnine – before setting out for a hunting excursion. At the end of a day's labor of shooting and skinning bison, they laced pieces of bison meat with the poison and left them for scavengers. Using this technique during a hunting trip in Kansas in 1879, Arthur Bill's outfit killed four hundred bison and fifty wolves in twelve days.[115] As the number of bison declined, "wolf-trapping" was so remunerative, the bison hunter George Simpson claimed, that some hunters concentrated on poisoning wolves during the winter.[116]

This ancillary destruction signaled the decline of southern plains bison hunting. Josiah Wright Mooar and his brother and partner John Wesley Mooar had the foresight to invest their profits from bison hunting in other business ventures. By 1877, they had purchased a cattle ranch and had begun to request cash payments from hide dealers, rather than leave their money on account against future purchases of supplies. By 1879, John Mooar reported to his mother, he and his brother had abandoned bison hunting: Josiah Wright was hauling supplies to the Colorado mines while John managed the ranch.[117] By 1880, Fort Griffin, Rath/Reynolds City, Buffalo Gap, and Hide Town were nearly deserted.[118] The bison hunters returned to work for the Santa Fe Railroad, went to the silver mines of Arizona, or moved north to pursue the bison in Montana and Wyoming.[119]

After the bison had been hunted out by 1883, the plains were strewn with their bones. A. M. Bede, a county judge from Fort Yates, North Dakota, recalled of his early days in the northern plains, "the country out here used to look like a charnel house with so many skulls staring at a man, and so many bones that

---

[112] Faulk, *Dodge City*, 32–33.
[113] H. B. Lovett, interview with L. F. Sheffy, 23 June 1934, Pampa, Texas, Panhandle-Plains Historical Society.
[114] Cator Family Papers, Panhandle-Plains Historical Society.
[115] Arthur C. Bill, "The Buffalo Hunt," Kansas State Historical Society.
[116] Simpson, interview with Haley, Canadian, Texas, 18 July 1926, Panhandle-Plains Historical Society. For the poisoning of wolves, see also Frank J. North Papers, State Historical Society of Nebraska; and Wentin Wilson, "Details of a Buffalo Hunt," Kansas State Historical Society.
[117] John W. Mooar, Fort Griffin, to his mother, Pownal, Vermont, 5 March 1879, John W. Mooar Papers, Southwest Collection, Texas Tech University, Lubbock, Texas.
[118] Dary, *Buffalo Book*, 114.
[119] Frank Collinson Papers, Panhandle-Plains Historical Society.

newcomers felt nervous, and, in some cases, could hardly plow the land."[120] L. C. Fouquet of Kansas also noted that the bones were "a nuisance to our breaking of the sod."[121] Poor homesteaders and Indians soon found a use for the bones, however, particularly during the drought years of the 1880s. They scavenged the plains for bison bones, which they sold to bone dealers for delivery to sugar refineries or fertilizer plants. It took one hundred skeletons to amass one ton of bones, which sold for $4 to $12. Although the scavengers did not become wealthy, the bone trade sustained many poor families. An early settler in North Dakota recalled that his family "hauled fourteen tons of buffalo bones.... I don't know how we would have lived if it had not been for the money we got that way."[122] Robert Wright of Dodge City believed that "if it had not been for the bone industry, many poor families would have suffered for the very necessaries of life. It looked like a wise dispensation of Providence."[123]

The bone trade was immense. In 1886, a resident of Dodge City saw a rick of bison bones a quarter of a mile long and as high as the bones could be thrown.[124] Bone dealers, such as the Northwestern Bone Syndicate of North Dakota, purchased thousands of tons of bones each year. Railroad companies annually shipped nearly five thousand boxcars of bones. The cargo was destined for the Michigan Carbon Works in Detroit and the Northwestern Fertilizer Company and Empire Carbon Works of East St. Louis, Illinois. (See Figure 5.5.) By the mid-1880s, the Michigan Carbon Works produced five thousand tons of bone black (a pigment) and four thousand tons of bone ash (a fertilizer) every year.[125] The bone trade was the final episode in the reduction of the most prominent resource of the plains to its salable parts and their incorporation into the American industrial economy.

The hide hunters' slaughter of the bison was a boon to Eastern tanneries, who had paid an average of $4.20 per hide in 1870, but were paying only $3.40 a decade later.[126] Moreover, the destruction of the bison accomplished the aims of the policymakers who sought to pacify the plains nomads. Once the hunters had eradicated the herds, the nomads had no recourse but to go to the reservations. Like poor Euroamerican farmers, the Indians gathered bison bones for sale to

---

[120] Bede, Fort Yates, North Dakota, to Edmund Seymour, New York City, 29 October 1919, American Bison Society Papers, Conservation Collection, Western History Department, Denver Public Library, Box 275, File 14.

[121] Fouquet, "Buffalo Days," *Collections of the Kansas Historical Society*, 16 (1923–25), 347.

[122] Quoted in LeRoy Barnett, "The Buffalo Bone Commerce on the Northern Plains," *North Dakota History*, 39 (Winter 1972), 23–41.

[123] Wright, *Dodge City*, 154.

[124] O. H. Simpson, Dodge City, Kansas, to the American Bison Society, 19 January 1934, American Bison Society Papers, Box 276, File 7. See also Joe Killough, interview with D. T. Leachman, 16 December 1945, Panhandle-Plains Historical Society.

[125] Barnett, "Buffalo Bone Commerce," 23–41. Bone black and bone ash are both produced by calcining bones, the former in closed vessels and the latter in open air.

[126] McMartin, *Hides, Hemlocks*, 88.

Figure 5.5. Bones at the Michigan Carbon Works, Detroit. Courtesy of the Detroit Public Library.

dealers. A reporter for *Harper's* magazine asked in 1893, "Do the Indians make a living gathering these bones?" Yes, replied a railroad inspector, "but it is a mercy that they can't eat bones. We were never able to control the savages until their supply of meat was cut off. We have had no trouble worth speaking of since 1883, however."[127] On the reservations, the Indians' existence was meager. They subsisted in part on government beef. At the Red Cloud Agency in western Nebraska, Oglala Sioux hunters prepared for the delivery of cattle from the government as they had once prepared for a summer bison hunt. As the cattle were released into the corral, the Indians set after them on horseback, slaughtering them in imitation – culturally rich but economically impoverished – of the communal bison hunts of the past.[128] Government authorities permitted the communal hunts to continue until 1897 when, in an effort to stifle this form of cultural expression, they built a slaughterhouse and began distributing butchered meat to the Indians. The Oglalas responded by setting fire to the slaughterhouse one night.[129]

The costs of the near-extermination of the bison were not shared equally. The plains nomads suffered the loss of their primary resource and consequently of most of their lands and economic autonomy. The benefits of the destruction of the bison were likewise inequitably distributed. Hunters transformed millions of bison into profitable commodities, yet merchants and industrialists appropriated most of the wealth generated by the destruction of the herds. While poor Euroamericans and Indians scavenged the plains for bison bones, industrialists added to their wealth by transforming the bones into fertilizer. In the plains, the slaughter left in its wake not wealth but poverty and misery.

Euroamericans did not slaughter millions of bison between 1870 and 1883 believing that nature provided an inexhaustible supply. Rather, they anticipated the extinction of the species. They regarded the disappearance of the herds as a triumph of civilization over savagery, because the extermination of the bison removed the nomads' primary resource and cleared the plains for Euroamericans. Hide hunters harbored little apprehension that the late nineteenth-century legal order might rein in the slaughter. Legal and extra-legal authorities in the nineteenth-century United States were the partisans of Euroamericans in their struggle to wrest control of resources from the Indians.[130]

The erosion of the nomads' hunting territory presaged the triumph – an

---

[127] Hamlin Russell, "The Story of the Buffalo," *Harper's New Monthly Magazine*, 86 (April 1893), 798. See also Mrs. Mae Tubb Dolcater, interview with Morris Dolby, 1 January 1948, Panhandle-Plains Historical Society.
[128] Collinson Papers, Panhandle-Plains Historical Society.
[129] James R. Walker, "The No Ears, Short Man, and Iron Crow Winter Counts," in *Lakota Society*, ed. Raymond J. DeMallie (Lincoln: University of Nebraska Press, 1982), 153.
[130] See McEvoy, *Fisherman's Problem*, 93–119.

inevitable triumph, in the minds of many contemporaries – of Euroamericans. The engine of the advancement of Euroamericans into the plains was the ability of an industrial society to destroy thoroughly the bison herds and thus deny their use to the nomads. A few lawmakers saw the destruction as deplorable – a violation of the treaties, a provocation to Indian war, and a profligate waste – but others saw it as a salient example of Euroamerican industriousness. The plains had remained a wilderness while in Indians hands, but hide hunters extracted extraordinary wealth from the grasslands – even if most of them failed to keep that wealth for themselves.

At the root of the failure to regulate bison hunting was the mid-century belief in economic competition. Everyone, Indian or Euroamerican included, was engaged in a race to exploit resources for individual gain. To reserve resources for anybody's exclusive use violated the competitive ideal; to reserve them for social outcasts such as Indians was unthinkable. Euroamericans waged a scorched-earth campaign against the Indians who impeded the expansion of industry. Yet the hide hunters' victory was hollow; when the campaign was over, most of the hunters found themselves no wealthier than before.

INTRODUCTION

Define, on the two-dimensional surface of the earth, lines across which motion is to be prevented, and you have one of the key themes of history. With a closed line (i.e., a curve enclosing a figure), and the prevention of motion from outside the line to its inside, you derive the idea of property. With the same line, and the prevention of motion from inside to outside, you derive the idea of prison. With an open line (i.e., a curve that does not enclose a figure), and the prevention of motion in either direction, you derive the idea of border. Properties, prisons, borders: it is through the prevention of motion that space enters history.

Abstract topological structures—closed, open lines—need to be implemented. Their physical (and social) implementation may vary. We may have absolute material barriers, whose function is to make motion impossible: such are walls, in aspiration. Or there may be more subtle obstacles, whose function is to make motion inconvenient and therefore undesirable: these, in general, are fences. Finally, there might be purely symbolic definitions of limits—a yellow line painted on the pavement—respected solely by virtue of the habits of social practice. Yet as with all other forms of coercion, even the symbolic definition of space relies ultimately on the potential presence of force (where there is a yellow line, there are usually also police nearby).

The ubiquitous presence of potential force is indeed a universal of history. Force, brute or refined, is what societies and histories are built of. Note, however, that with the prevention of motion, force —in the most literal sense, of applying physical pressure to bodies— assumes a special kind of necessity. Quite simply, being in a place is something you do with your *body*—nothing else—and therefore, to prevent your motion from one place to another, your body must be affected. The history of the prevention of motion is therefore a history of force upon bodies: a history of violence and pain.

Facilitation of motion is another important theme of history. In this book, I will often have occasion to mention not only dividing lines but also connecting lines: sea-lanes, trails, railroads. It should be seen, however, that the prevention of motion is in a sense more fundamental than the facilitation of motion. A train is worthless unless you can prevent some people—those who did not buy your tickets—from boarding it. Like all property, the train becomes valuable only when access to it can be controlled, and so the system of the railroad—lines that connect points—is anchored by the system of *stations*, buildings whose walled lines enclose space and control motion. A world where the railroad exists without the station is unthinkable, because without control over motion, value cannot be formed. Value arises from lines of division—even when they happen to enclose lines of connection. To understand history and its motions, then, we must first understand the history of the prevention of motion.

This book follows one of the major threads of this history. I show the conditions for the invention and spread of a simple but highly significant technology: barbed wire. Starting with a description of its origins in the colonization of the American West, I move on to describe its eventual role in modern warfare, and then in the modern forms of human repression, offering finally some remarks concerning the general lessons that may be derived from the growth of barbed wire. Throughout the book, I glance beyond barbed wire to the space it has enclosed. Around the strand of history made of barbed wire, I weave a chapter of modernity. Barbed wire allows us to see a more fundamental ecological equation, whose main protagonists are flesh and iron. Here is how modernity unfolded: as iron (and, most important, steel) became increasingly inexpensive and widespread, it was used to control motion and space, on a massive scale, exploiting its capacity for mass production and its power of violence over flesh. This massive control over space was the defining characteristic of a certain period of history: the eighty years from 1874 to 1954—from the invention of barbed wire to the downgrading of the Gulag. Throughout this period, barbed wire constructions were at the forefront of the major events of world history. This was not an accident; barbed wire was what this period required. This book tries to explain why. Thus the book is about what may be considered the age of barbed wire: the period of the coming of modernity.

This history took place precisely at the level of *flesh*, cutting across geographic as well as biological boundaries. It was not humanity alone that experienced barbed wire. The tool was created to control animals by inflicting pain on them. The enormous sweep of barbed wire through history—ranging from agriculture to warfare and human repression, encompassing the globe—is due to the simple and unchanging equation of flesh and iron. The first must yield to the second, followed by the inevitability of pain. The history of violence and pain crosses species, and so, as a consequence, did the history of modernity.

It is only by considering reality at this level, going beyond humans alone, that history can make sense. Indeed, although much has been written about some technical aspects of the invention of barbed wire, the particular thread followed in this book seems hardly to have been noticed at all. Some authors have written about barbed wire in agriculture, many more about concentration camps, but few have even mentioned both in the same breath. This is precisely what needs to be done if we wish to understand either concentration camps or agriculture. Both belong to the same world and follow the same history. This book is largely about the environment—literally—that gave rise to concentration camps: as it were, an environmental history of Auschwitz. It thus has to start where environmental history does, in the encounters between humans and other animals. For animals are always part of the social picture; their flesh, suffering and consumed, motivates human history itself. When we set out to offer a history that mentions animals, we should understand that the history of animals is not merely an appendix, a note we should add because it is missing in our present traditional, human-focused history. Rather, the history of animals is part and parcel of history—that reality where all is inextricably tied together, humans, animals, and their shared material world.

1 EXPANSION

The American West and the Invention of Barbed Wire

Your first question is simple: when and where was barbed wire invented? Let us start with the simple answer: barbed wire was invented in 1874, for use on the American Great Plains. Let me be more precise so that we may begin looking for the essence of the tool. Its goal was to prevent the motion of cows; its function relied on violence; its success depended on deployment on a vast scale.

The question may be restated: why would America need, in 1874, to prevent the motion of cows on the Great Plains,[1] and why would it do so through violence deployed on a vast scale? Now we reach some difficult questions whose answers might reveal the nature of barbed wire.

Barbed wire was created as a result of a special kind of colonization taking place in the American West. This colonization had two features that, combined, set it apart from earlier colonizing episodes. First, it was new in terms of space: an entire landmass was to be exploited (and not merely some selected points on it). Second, it was new in terms of time: the colonization was to take place very rapidly. Earlier human expansions on similar scales had taken generations, but this one was to take no more than a few years. There were precedents for massive colonizations, and there were precedents for rapid colonizations, but there was no precedent for a colonization that was simultaneously massive *and* rapid. Thus a new way of control over space was called for: one based on violence deployed on a vast scale.

In this chapter's first section, "Unpacking the Louisiana Purchase," I consider the rise of the cow on the plains, leading to the immediate background to the invention of barbed wire. Section 2, "How to Fence a Cow," describes how cows were controlled by barbed wire. We see how space and its animals were suddenly brought under control following the introduction of barbed wire. In section 3, "How to Fence the World," I consider the shape of the industry following its globalization. The problem of control over animals was universal, and it resulted quickly in a system, based on the one used in the American Northeast, reaching around the globe. The following two chapters will trace out the consequences of this globalization.

1. UNPACKING THE LOUISIANA PURCHASE

No one quite knew what to do with Louisiana—nor, indeed, where precisely it was. The French colonial claim of that name, dating from 1699, more or less coincided with the borders of his Great Plains (though the term was not used), and the borders were left intentionally vague so as to leave room for hypothetical future expansion. In truth, control over space was a mere act of cartography, and to name this particular space after Louis XIV was about as practical as Galileo naming Jupiter's moons after the Medici. The territory was so little known that one could not define it with any prominent features of the topography. Thus authors were reduced to the indefinites of symbolic space. Take, for instance, Du Pratz in 1763: Louisiana was "that part of North America which is bounded on the south by the Gulf of Mexico, on the East by Carolina . . . and by a part of Canada; on the west by New Mexico, and on the north by parts of Canada, in part it extends without assignable bounds to the terra incognita adjoining Hudson Bay."² Exactly the same ambiguity applied to all the other parameters—Carolina, New Mexico, Canada. The equation was not meant to be solved. Anyway, by 1763 the question was moot; France was defeated, and her possessions were divided, east of the Mississippi (i.e., Quebec) to Britain, west of the Mississippi (i.e., Louisiana) to Spain. This defined the eastern border and made the western border unimportant. (Louisiana was now merely the name of another Spanish colony, just like Texas to its West.) Not that any of this really mattered to the Indians or the bison—the true inhabitants of the plains—whose life was dominated by another, more real geography of the spread of horses, guns, and smallpox. In Europe, however, the spaces of the American continent took on dramatic dimensions. Returning in 1800 from his failed Egyptian expedition—where he tried to derail the British Empire through the East—Napoleon decided to attack from the West. Napoleonic pressure returned Louisiana to the French, a base from which to disrupt British power in Canada and in the Caribbean. No more luck for Bonaparte here than in Egypt, though; a slave revolt in Haiti made the French position in the Caribbean tenuous. Meanwhile American diplomats, worried about the presence of the bellicose French at the mouth of the Mississippi, inquired whether New Orleans could be leased. Napoleon, quickly reconsidering his position, retroactively made the entire operation into a real estate investment. He offered the territory to the Americans, all for $15 million. The Americans, by no means naive themselves, then obtained excellent conditions of payment, the entire sum paid in American public debt. In 1803, when the transaction was made, almost nothing changed hands. Napoleon sold the Americans the promise of space and was paid with the promise of money.³

Absurd as they might seem, early-day colonial claims were not irrational. Inside the phantom territories staked by such claims, real interests were protected. This was the trader's colonialism that made its big profits not by covering areas but by connecting points: a plantation, a mine, a market, a port. The spatial commodity exploited is distance. Sumatra is very distant from Italy, Peru is very distant from China, Jamaica is very distant from France. Pepper, silver, and sugar, crossing such distances, multiply enormously in value. By gaining control over this network of shipping from procurement to consumption, one obtains a tremendous source of profit and power.⁴ In this type of colonialism, then, perimeters of influence were meant to enclose not a space but a series of points, and details of border and control over space were irrelevant.

Distance would always remain valuable, and it still is: the value of the sweatshop is a function of its distance from its clientele. The geographic distance allows a vast separation between the extremes of poverty and affluence. Connecting these two extremes is crucial to the contemporary world, as it has always been to colonialism. However, in the nineteenth century, a new kind of colonialism emerged: not the trader's but the *investor's* colonialism. The investor does more than connect: he invests and develops, turning as much as possible of a land into usable resources. This new colonialism, then, would be the investor's colonialism—based on the profits to be made out of intensive production on a vast scale. The spatial commodity for this colonialism is not distance alone but also area itself. Thus borders would now be defined, and their interiors thoroughly controlled. And this was to happen now, following America's acquisition of the West. In the nineteenth century, America led the way to the world in making the transition from the trader's to the investor's colonialism. In taking up the Louisiana claim, America entered, without knowing it, not only a new space but, more important, a new way of handling space. This would ensue in ever-widening cycles of violence.

It started with Texas. A Spanish territory in 1803, Texas came to be part of the newly independent Mexican state in 1822. Its many colonists from the United States took a dim view of the antislavery position of the Mexican government. They fought for their freedom to own slaves—gradually drawing the United States itself into their protection, which, in the brief war of 1847, finally led to the United States being in possession of the West.

At this point, the countdown began for the American Civil War. The issue is this: Land, alone, does nothing for humans. It has to be used in some definite way: cultivated; grazed by animals; mined; built. Each land use determines a different ecology and thus a different society. To reach for land is to try to extend a certain social order into it to the exclusion of others. Decisions are painful, especially when different social orders coexist already. To open new lands is therefore to open old wounds; America's wound, of course, was slavery. This land use was based on extensive agriculture in large fields, poor in technology, rich in the unskilled, coerced labor of draft animals and enslaved humans. It was most profitable in the global products of classical trader's colonialism: sugar, tobacco, cotton. It assumed little investment and much transportation. Contrast this to the land use of the farmstead, where a paterfamilias would govern a large family and its livestock to make a living from a land intensively cultivated, partly for internal consumption, partly for the sale, to nearby urban centers, of high-quality produce. This is based on cheaper transportation, but more intensive investment. From the census of 1860, we can take the following two questions: (A) the numbers of acres of improved land (irrigated, fenced, etc.) on the farms; and (B) the number of acres of unimproved land on the farms. A vote for Lincoln was directly correlated to the ratio of A to B. Connecticut had nearly three improved acres of land for each unimproved acre; South Carolina had nearly three unimproved acres of land for each improved acre. This was the divide defining the Civil War.

We can mark the big divide as follows: between a northern country, where fields were controlled by the intensive use of the fence, and a southern country, where fields were controlled by the intensive use of the whip. At issue, in other words, was not just the moral sentiment of abolitionism—which, it is only fair to say, did greatly move many Americans—but the realities of control. The North wanted to see an America with acre after acre of improved land supporting both families and urban centers, all inspired by the intensive economy of the northern American town. The plantation owners of the South wanted to prevent just this outcome of an America governed by the cities of the North. The South needed to grow, demographically, just as the North was growing, and the South needed new slave states, if only to have new slave state senators.

Now the settling of the plains themselves gained a new urgency. It was progressing apace; the bison were retreating, and the railroad was beginning to send its branches west of Chicago. The continental rail project—to connect California with the East, gathering all the West along the way—exacerbated the sectional strife and got stymied by it. The decision about a route for this train would be, symbolically as well as practically, a decision about which part of the nation had first claim to the West. Hence no decision could be made. Nor could any explicit political decision be made for the settlement of the plains. Yet throughout this all, American agricultural practices were brought into Kansas and the Indians. In Kansas, Southern and Northern farmers faced one another in what became throughout the 1850s a bloody skirmish, a prelude to the Civil War. Simultaneously, the West was being integrated into the East, and the South was breaking away from the North—and the two processes were one.

To repeat: there was a fundamental asymmetry between North and South. Northern farms were outposts of agricultural production sent out by an urban, industrial economy; this is ultimately why intensive farming made sense for them. Southern farms were all the South really had. Like the entire Caribbean area, the South was all, essentially, no more than an outpost of Europe. The sectional divide was thus a relic of the trader's colonialism, of a time when America was made up of discrete entities serving separate European functions. With its railroads, with the explosion of its urban life, the North was now ready to become its own center, and the center for the entire continent. More than this, the North was ready to become the center of the continent *as a continent*, the entirety of its land being developed for the support of Northern cities. This was why Southerners felt threatened, and why they lost. When Lincoln came to office in 1861—a president whom Southerners perceived, somewhat falsely, to be a Free-Soiler—and when the South finally seceded, the war was fought not merely to resume an American system. The war, instead, *created* such a system. Now it was to be, for the first time,

a single structure, with a single center based in America itself, on the northern Atlantic seaboard—all intensively developed. The war started because there was a West to incorporate, and it ended with the West—as well as the South—both incorporated into capitalist America.

In Lincoln's Congress—the Southern filibusterers now having seceded—all the gridlocked issues of the 1850s were pushed into motion. The main goal was to develop land in the West; the main tool the government had at its disposal was land in the West. Thus the curious nature of the legislation, offering uncharted lands to those who would chart them. The Pacific Railroad Act established a northern route for the railroad, offering its developers, as incentive, 6,400 acres of western land (more would be decreed in the future) for each mile constructed. Meanwhile, toward the foundation of state colleges, the Morrill Act gave western land at the rate of 30,000 acres for every senator and congressman each state had. These colleges, let us remember, were primarily supposed to produce agricultural experts—which, in the early years, is what they largely did. Funded by the intensive cultivation of land, their intellectual production served to intensify agriculture further. Finally, the Homestead Act was the crucial legislation that set out the basic form of settlement for the West. The act promised each individual settler, following five years of residence and improvement, 160 acres of land. This envisaged small-scale, intensive family farming. The railroad, agricultural science, Northern farming families—all were expected to replicate soon, on the Great Plains, the economic achievement of the North.

This was all enacted in 1862 in Washington, D.C., while not far off, Americans were dying in numbers—and ways—unimagined. The Civil War was the fourth cycle of violence unleashed by Louisiana, following Texas, Mexico, and Kansas, but nothing had prepared for what happened now. It was as shattering to contemporary Americans as World War I would later be to Europe. It was strange and frightening; while warring, war itself was changing. No one knew iron could wreak such havoc. Ironclads, introduced in 1861 by the South and soon mass-produced by the North, made wooden military ships obsolete overnight. Railroads allowed the concentration, never seen before, of hundreds of thousands of soldiers. Rifles—an invention assembled together during the 1850s—changed the space of battle itself. If you impart spin to a bullet by shooting it through a rifled, or spiral-grooved, barrel, it gains in precision and thus in effective range. The smoothbore musket had a range of not much more than a hundred yards; the rifle had one of about six hundred, covering a space thirty-six times greater. From 1862 onward, the hundreds of thousands of soldiers amassed by the railroad carried with them rifles instead of muskets. Iron made battles larger: the rifle made the field of killing greater, and the railroad enlarged its reach in terms of human population. A soldier could be drafted in Boston, within a few weeks arrive in Pennsylvania, and there become tangled in an area of tens of square miles of unremitting violence—the worst of them all, the field of Gettysburg, where more than 50,000 were killed over three days' fighting in July 1863. Death was agonizing; rifles were at an interim stage of technology, forceful enough to get the bullet inside the body even at long range, but not quite forceful enough (as twentieth-century guns would be) for the bullet to exit the body following impact. Civil War bullets typically rested inside the flesh, ensuring inflammation and, in most cases, painful death.[5] All in all, more than 600,000 Americans died in the four years of the war. The brutality of the frontier skirmish—the Indian wars, Texas, Mexico City, Kansas—returned, magnified many times over, to the centers of American civilization. I will return to this dialectic of frontier and center—the brutality of the first returning to haunt the latter—in the next chapter.

Not that the American frontier skirmishes ever stopped. The Civil War had its Indian War built into it. Indeed, some tribes made the wrong tactical decision, siding with the South—particularly in the Indian Territory. This was very convenient for the North, as ultimately it would allow the federal government to discontinue the grant of *any* territory to the Native Americans. But the same was true everywhere. The skills, the brutal attitudes, and the technologies developed in the Civil War were seen in the West as well. Even as the Civil War was still raging, rifles shot more bison than humans.

These were the cycles of violence: from the Texans' war against Mexicans, through the Mexican-American War, and then through the North-South skirmish, particularly at Kansas, came the Civil War itself; and this led immediately to further cycles of violence, aimed now at the Indian and the bison. In November 1864, General Sherman was marching from Atlanta to the sea, everywhere proclaiming the cause of liberty. Just then, far to the West, the Cheyenne Indians were invited by the American settlers to come to Sand Creek, Colorado. The Cheyenne were promised that they could hunt there, but on November 29 they were hunted themselves. Local Colorado militia forces surprised the Cheyenne in their tents, and all were killed—hundreds of men, women, and children. Skin cut off a dead body had an enormous fascination for the killers of the West, and the scalps of Cheyennes were now displayed, to applause, in Denver's public theater.[6]

Such excesses were indeed less common, and an outcry took place when news reached further east. But America did not really have an alternative Indian policy. To start with, the main piece of official policy were forts garrisoned across the West to protect the growing railroad and the concomitant agricultural settlements. In topological terms, then, the Great Plains were a plane surface, across which points (garrisons and settlements) were connected by lines (railways and trails), the surface as such still belonging, in a sense, to the Indian and the bison. Precisely this topology was to be changed. The bison—the basis for the Indians' way of life—were being finished, and the Indians were urged to settle down, to get out of the way. Instead of Euro-Americans being confined to points on the surface, the Indians were to be reduced to their points—the reservations—the entire surface now becoming European. This was the enlightened alternative to Sand Creek. Indians, on the whole, realized they had no other option, but many resisted. They had moved to the plains from the East, generations ago, because their agriculture was failing under European pressure; they had taken to hunting because, with their resources, successful agriculture on the plains was impossible. They suspected they were being condemned to a life of destitution, and they were right. But all their courage and equestrian skills notwithstanding, the Indians had no chance. With the typical advantages of guerrilla fighters—better mobility, knowledge of the land—surprise and individual successes were always possible, most spectacularly at Little Bighorn, when on June 25, 1876, Colonel Custer was caught and killed with his force of 220 soldiers. But in fact, these were already the last moments of Indian resistance. They had nothing to roam the plains for. The bison were now dead, replaced by railroads and farmers. As the Indians retreated to their pitiful reservations, the cow began its trek north of Texas, eventually to introduce there an economy based in Chicago. And this, finally, was the culmination of American history in the nineteenth century. Texas led to Mexico, which led to Kansas, which led to the Civil War, upon whose conclusion America could move on to destroy the Indian and the bison. The final act in the subjugation of the West was under way: the transition from bison to cow.[7] This was the immediate consequence of the Civil War: the West was opened for America—and America had filled it with cows.

We are getting near the invention of barbed wire, then. So let us focus our attention on western cows, at the moment when they replace the bison. Was this, in reality, a deep transformation at all? The answer is complicated. At first glance, the new order could be said to be no more than a shift of species and of race: bison replaced by cows, Indians replaced by Euro-Americans. Neither shift, in itself, involved, at first glance, a dramatic change.

Take first the animals. The Texas longhorn cow, instead of the herds of wild bison, now roamed the plains. We should not be misled. When one thinks of a cow, what comes to mind are, perhaps, dairy cows seen in European fields—heavily bred and disciplined so as to produce a breed as docile as a spaniel. But the ancestors of these cows had gone wild after being brought to America by Spanish colonizers. The same happened to many domesticated species brought to the New World. Animals, let loose on a new continent, outgrew their European past, indeed, their European masters. The local ecology had little to resist the new species, and a few escapees would be enough to establish a huge population, gradually shedding its domesticated habits.[8] Beyond the limited domain of European settlement and domestication, a penumbra of feralized animals could be seen on the American continent. Here were wild horses—as many as two million of them—famously contributing to the last stage of Indian life.[9] So, to a lesser extent, were wild cows. In the 1870s, they were just being brought back to the fold, and the Texan breed was still remarkable in its ferocity. Nearly self-sufficient, they were thus not totally unlike the bison that they had replaced.

As for the Indians—for the last century subsisting almost exclusively by hunting the bison—they too were replaced by a breed not quite unlike them. Euro-American men, mounted on horses, gathered and herded the cows, roaming the same plains as the Indians did, following the same constants of grass and water, living in similarly small bands with little attachment to settled community. Life on the plains, then, did not change much.

The essential ecological structure was in a sense preserved as well. The sun's energy was stored up by grass. The grass was consumed by vast numbers of large bovines. These in turn were herded and killed by small bands of humans. At first, perhaps, not quite as many as the bison; the bison population is now estimated to have peaked at about 30 million near the beginning of the nineteenth century, while cattle numbered perhaps over 11 million by 1880. But then again, the rise of the cow came after a long period of degradation, as overroaming, human impact, and ecological catastrophes gradually reduced the capacity of the plains to carry bovines. Taking a longer view, we can say that the bovine population (i.e., either bison or cows) started from around 30 million at the beginning of the nineteenth century, collapsed to perhaps 15 million in the 1860s before the final onslaught on the bison herd, then bottomed out in 1880 at 11 million before climbing back to nearly 24 million by 1900. The death of the bison was in a sense merely a crisis of transition from one bovine ecology to another.¹⁰

Bovines, far more numerous than any other mammals, continued to dominate life on the plains. They were dominant also in the sense that they governed space, at least locally. Just as the bison did, cows roamed freely most of the time—and just as the bison did, cows did it all under the surveillance of small bands of humans. Finally (and here is the essence of the continuity) the cows, like the bison before them, accounted for the presence of the humans. Everything about human life on the plains was built around the protection of bovines for the sake of their future killing, just as it had been since the start of the Indian hunting experience on the plains. In a sense, the American West had to start from somewhere, so it started from where the Indians left off. There was nothing better to be done with the land.

Below the ecological continuities, however, ran deep differences, most obvious in the nature of the killing. The basic structure of the history of the Great Plains was the evolution of methods for killing bovines. In fact, killing a bovine is not an easy thing to do. A bison, in particular, is a swift, agile animal. Of course, the bison did not evolve to be protected from humans, but it had enough experience with wolves and other mammal predators to teach it caution. Prehistoric Indians could hardly face a bison and try to kill it; it would, quite sensibly, run away. This was the bison's mistake: it should, of course, have turned around and tried to ram the Indian, but the bison never realized how much weaker humans are than wolves. Thus the Indians could elaborate their method of killing. It worked like this. First, the hunt was at the level of bands—a band of Indians gathered together against a band of bison (single bison or small groups would not be affected by the method I describe). The bison would be frightened, literally, out of their wits. The humans egging the bison on would gradually herd them along a predetermined route. There they reached a precipice (the plains, in fact, do have their hills). The bison, being closely packed, could not change direction at the last moment. Most if not all would fall over the brink, which, even if not very

high, would suffice to shock them so that they could be done away with by the band. Notice that all tribe members participated in the exercise, which was almost pastoral, rather than hunting, in nature.¹¹ Then, in early historical times, dramatic changes took place, and the Indian hunting method changed completely. With the horse—rapidly made available on the plains during the eighteenth century—equestrian hunters could now outrun the bison and kill it from horseback. Note the advantage: killing was possible all year long, not only during the rutting season (when bison would form their great bands). Note also that, now being more capital intensive, so to speak, killing a bison became more specialized and involved a division of labor. The Indian women, with a lesser contribution to food procurement, developed the specialty of making robes from bison hides into something of a manufacture business. Soon Euro-American merchants, reaching up the river on the Mississippi, would prompt the Indians to kill bison specifically for the purpose of robe making.¹² Finally there came better guns, and in particular the much more precise rifles, invented in the 1850s and used, as we have seen, to great effect in killing humans, too. Armed with these, Euro-Americans overwhelmed the bison—already decimated by Indian overkill. Now the killing of the bison was more capital intensive (you needed to own a rifle), but almost labor free. There was no problem whatsoever in getting the bison into rifle range, so that the plains practically became bison-killing factories, with rifles for machines. The hunt peaked in 1872, and the plains were practically clean of bison by 1883; according to one estimate, more than 5.5 million bison were killed in the peak years of the early 1870s alone.¹³ At this point, division of labor as well as capital investment went one step further. The Euro-Americans were killing bison not to eat their meat but to transport the unprocessed hides east. The bison, killed by the products of American machinery, were further processed by this machinery—and then became part of it. The bison hide was processed by the American tanning industry to produce, in particular, the strong belts required for running factory machines.¹⁴

The cow brought this process to its logical end. We have seen how able, if dwindling, natural resource. The handling of cows, finally, represented a fully capitalist economy, with sharp division of labor. Killing was now made fully calculated and economical. There was more revenue in a cow than in a bison, but less of a profit margin: more thought, therefore, would go into the cow's killing. The cow economy—as well as the cow ecology—would not have the simplicity of structure that the plains had at the times of the bison.

the bison, in the final stage of its existence, stopped being consumed or processed on the plains; the cow was not even *killed* on the plains. The plains merely transported the cows now and gave them whatever meager nourishment would sustain them through the process. A cow would typically begin its life in Texas; herded north, it would roam under the guidance of humans somewhere in the plains, then be herded again eastward (sometimes by rail), often to be better fed and cared for there, briefly, nearer a major center of slaughter (Chicago itself, or some urban center further east). This last stage of care was necessary because of the immense hardship the cows had just been through. Walking the entire American Midwest, often under inclement weather and in inhospitable terrain, was an experience reflected by the animals' physical state—and so in their commercial value. To make them more profitable, therefore, they were allotted a brief period of comfort before death, as if to compensate for the months and years of deprivation. Finally, however, the animal would be brought into a city to be killed there, its carcass processed and then finally consumed. During this process, many humans would be involved: usually more than one group of cowmen and farmers, freight train personnel and retailers, farmers again, and then a butcher, leading finally to the consumer.

This new complexity had two aspects. First, the biological dominance of cows in respect to other nonhuman species would soon be challenged: land would be used not only for the feeding of cows but directly for agriculture. Second, and related, the relationship of cows with humans was much more one-sided than bison-Indian relations had ever been. Of course, the Indians were dominant enough relative to the bison; they could kill hundreds and hundreds of them with great ease. But, after all, the bison was also the great imponderable of Indian life on the plains, the beast whose numbers and appearances were to be determined by forces beyond human control. For capitalist America, nothing was supposed to be beyond human control.

What is control over animals? This has two senses, a human *gain*, and an animal *deprivation*. First, animals are under control as humans gain power over them—most importantly, as humans gain control over animals' biological cycle (procreation, growth, and death). Such control transforms biological patterns into marketable commodities: this is the essence of domestication. Second, animals are under control when they are deprived of their powers of activity. To survive, an animal must develop a certain control over its environment: it can move, trace food, graze or kill. To complete the control over the animal—to reduce it to a mere passive member in a fundamentally human society—is the other side of control over animals. Where they cannot be domesticated, then—that is, when they cannot be reduced to mere passive members in an otherwise human society—animals have to be kept away or destroyed. When all animals have been either subdued or destroyed, a share of land has been cleared from animal power

Horse, steamboat, gun, railroad—as each tool of control over space reached the plains, a further step was made toward capitalism. Now, finally, capitalism was reached. The prehistoric bison hunt represented a precapitalist economy, with the killing limited by humans' precarious hold over their environment. The historic bison hunt by Indians represented the unstable interface of capitalist and precapitalist economies. With relatively little division of labor overcame reality. Extremely vulnerable in this exchange, becoming ever more dependent on American merchants, the Indians were driven to overkill and to ruin the basis for their way of life. At this point, with hardly a life left in the bison herd, Euro-American hunters, representatives of a more sophisticated capitalist system, came to exploit what was for them merely a valu-

and brought fully under human control. The fact is, control over animals is rather like control over humans: you can either make them do what you like them to do or else get them out of the way. This is how societies are made: human societies as well as the larger, multispecies societies that humans have created.

Now, large animals living in large social groups—the kind, that is, of most direct value to domestication—combine the considerable force of each member to create, in their herds, a powerful social organism. Thus they pose an especially difficult task in trying to bring them into human society. Facing the bison, the human problem was stark, and the solution adopted by Europeans was very simple, that of extermination.

But even with fully domesticated animals, it takes a considerable amount of effort to set up a constant counterforce so as to keep the animals at bay. Even domesticated animals, after all, are still wild. And so they would, unless specifically controlled, go where they want, eat what they want. Hence the problem of subduing the cows on the plains. It was to solve this problem that barbed wire was invented.

2. HOW TO FENCE A COW

All those land grants—to the railroad, to the state colleges, to the homestead, to the many Louisiana Purchases. When you got there, there was not much to it. America evolved through the experience of the Atlantic, the Gulf, the Great Lakes, the Mississippi. Farmers built their life based on the expectations of copious rainfall and its attendant vegetation. Now they got to plains that evolved through the experience of aridity—mostly less than sixteen inches of rain per year. (Boston has over forty inches; New Orleans has over sixty.) Who would live there? Grass, bovines, wolves and other predators, humans foremost among those other predators. Grass could survive on little, unpredictable precipitation; bovines could survive on grass. Wolves and humans could then survive on bovines.[15]

From the point of view of the individual cattle ranger, that was just fine. The West may have been won by the North, but the immediate gain was made, once again, by Texas Southerners. As we have seen, the immediate aftermath of the killing of the bison was the herding north, from Texas, of the longhorn cow. In one respect already, this was part of capitalist America—the cow was to be killed in Chicago or further east. But in other ways, the practices of the range were simply transmitted north.

This explains the continuity with the bison-hunting practices described in this chapter's first section. In other words, the West now had a range, not ranch, business. Do not be confused. In contemporary agriculture—which tends to be, strictly speaking, a ranch business—the terms "range" or "open range" came to have the more narrow meaning of any animal raising that does not involve strict imprisonment inside a building. In the original sense, the distinction between "range" and "ranch" was different. A ranch is an enclosed piece of land; the range is space, unlimited. Originally, Texas cows were fed off the land and moved through it, all throughout the plains—just as the bison did. Control over parceled units of land—the essence of the land grants—was at first out of tune with the actual economy. The economic value of cows resided in their self-reliant properties—they found their own nourishment, and this meant, especially in the difficult conditions of the Great Plains, that they had to operate in open space. The profitability of animals was, at this stage, partly a result of animals acting independently, exercising their powers. A ranger has the animal not merely to be killed eventually but also to do the work for him: the ranger does not look for food and water for the cow; instead, it is the cow herself who seeks those resources. The range industry makes its profits by combining the killing of animals with their forced labor. (This, we should note, was hard labor, under very harsh conditions.) To reduce the motion of cows, then, is to reduce their labor and thus to take away from the owner's sources of profit. The truth was, there were so few resources on the plains that settling anywhere in particular, at first, made little sense. Better to move on with your cows, finding grass and water along the way. As the Indians were being consigned to their reservations, Texans were taking up a quasi-nomadic form of existence. Life was endless motion, and human survival was impossible without the horse.

Nor, indeed, would there be any compelling reason for an owner of cows to fence them so as to gain control over them. Not only did cows manage to survive on their own; they could also be relatively easily collected for marketing by small numbers of humans on horseback. Within the arid plains, river valleys—7 percent of the land—were the only space that mattered. The promise of an open plain actually reduced to the reality of branching rivers, on which, historically, life depended. Of course, cows might wander off, but one did not need constantly to inspect each of them individually. Control could be maintained in other ways: the river would determine the areas where cows could roam. They were boxed by the climate and geography of the plains. This way, herds would be assigned separate spaces along the banks of rivers. For practical purposes, a river's bank does not have its space open in all directions. Inland, away from the river, thirst blocks the motion of cows; the river itself blocks motion on the other direction. The rectangle along the riverbank has therefore only its two short edges open. All you need is to patrol these two edges.

Most important, you rely on the practices of the cow itself. This is the principle of domestication: study the habits of an animal and use them against it. The cows could become free from humans, but they were the captives of their habits. They were conditioned to protect themselves against predators by forming into close herds. Their gregarious habits are precisely what humans exploit. Cows just will not disperse. Had some herd realized in 1866 what it was up against, it could have made the rational choice and dispersed in all directions. No amount of cowboy skill would have been able to collect all the cows, and those that were left on the range would have had, at least, a sporting chance against the occasional wolf. But the cows never realized this; they kept going together, assuming that this was—as it had been thousands of years earlier—in their best interest.

Hence moving cows over long distances is a fairly simple task. The mounted humans who controlled the herds—frightening them all the way up to Chicago—kept an eye on them not so much to prevent them from running away but rather to prevent other predators from taking away the prize. Control over the cow itself was easy; this, after all, is why the animal was domesticated in the first place. No need for fencing, then, as far as the cow itself was concerned.

The threat of other humans was a special problem, of course, but once again, the division of land was not necessary for this purpose either. The goal for this type of economy was to establish control not over land directly but over the cows on it. Instead of marking the land, it was more rational to mark the cows. Thus, to guard against theft, owners branded their cows—an ancient practice applied systematically in the West.

For humans, of course, symbols are more than just practical tools; they embody culture. Branding was, and still is, a major component of the culture of the West. Ranches, for the last century, have often been named after their brands; pride is taken in mastering this symbolic system that defines human control over space and over animals. Ranchers will show off their ability to recognize the many symbols invented. The ritual is still central to ranching life: tying the animals' legs tightly together; setting a fire; carefully heating the branding iron (large, so as to make an articulate, clearly visible mark); then applying the iron until—and well after—the flesh of the animal literally burns. As put by Arnold and Hale (western authors writing in 1940), "There is an acrid odor, strong, repulsive . . . [the animal] will go BAWR-R-R-R, its eyes will bulge alarmingly, its mouth will slaver, and its nose will snort." (At this stage, typically, a bull will be castrated, and in many cases, a cut will be made in the ear as a further symbolic mark.)

A complicated ritual: as the same authors note, "[the inventor of branding] could hardly have suspected how much fun and interest would eventually center around cattle-branding."[16] The entire practice is usefully compared with the Indian correlate. Indians marked bison by tail tying; that is, the tails of killed bison were tied to make a claim to their carcass. Crucially, we see that for the Indians, the bison became property only *after* its killing. It was only then that the bison made the passage from nature to human society. The cow, on the other hand, was a property—indeed, a commodity—even while alive. It would be branded early in its life.

In the special case of calves born free, branding was the moment when a cow passed from nature to culture. A feral calf caught on the Texas plains, unbranded and technically called a "maverick," would be branded and thus made a commodity. Once again, a comparison is called for: we are reminded of the practice of branding runaway slaves, as punishment and as a practical measure of making sure that slaves—that particular kind of commodity—would not revert into their natural, free state. In short, in the late 1860s, as Texans finally desisted from the branding of slaves, they applied themselves with ever greater enthusiasm to the branding of cows. Sometimes whole herds would be collected through "mavericking" (the technical term for hunting and branding wild calves). Such herds would gain their marketable value by being herded north, as far as Chicago. This was a fortune that required, as investment, nothing more than motion and violence.

Violence, of course, was everywhere in the West. Central control did not yet extend across the land as a whole but was still limited to the network of military garrisons. Beyond that, power was wielded by small, mounted bands, ready to kill: the same kind of people who had fought for Texan slavery against Mexico, and then against Indians, or against each other in the Civil War. The habits of violence were endemic to the land. The combination of violence and motion, after all, is what made the West so cinematic.

But the West was even more interesting than that. Its myth was based not merely on violence and motion in the raw but on another, more subtle encounter: that between violence and motion, on the one hand, and civilization, on the other. This myth is fully based on reality: the North had won a war designed to make the West into an extension of its prosperity based on the prudence of farmers. It had also won this war through the experience of violence and lawlessness. Hence the liminal character of the West.

Consider, for example, James Butler Hickok, at one of the mythmaking moments of the West. The year was 1871, and Wild Bill, as Hickok was popularly called, was employed by the city of Abilene, Kansas, to act as its marshal. Abilene had been made practically overnight by the Illinois cattle trade when, in 1867, the rail terminus to Chicago opened there. The economy was booming: the cattle needed pens, attracting settled farmers; the cattle drivers looked for gambling and sex, attracting an altogether different kind of population. This led to the division of the city, neatly marked by the railroad tracks, between law and lawlessness. On one side was a midwestern small town transplanted further west; on the other side was the demimonde of a border town. Wild Bill was hired to prevent any spillover across the tracks. He was considered successful, but he was also deeply resented by the Texan herdsmen, particularly because of his past (born in Illinois, his adult life was divided between fighting Indians and fighting Southerners). Toward the end of the cow season, on October 5, he was passing the Alamo Saloon when a shot was fired, narrowly missing him. The Texan Philip Coe, pistol in hand, explained that "he shot a stray dog," and then fired again at Wild Bill, who immediately shot Coe twice in the stomach—as well as fatally shooting a bystander who came to help. This turned out to be Wild Bill's friend Michael Williams. Grief-stricken, Wild Bill set out to chase the Texans from town; a marked man, he lived now in constant fear for his life. This is the stuff western myth is made of; but consider the denouement of the combat. Abilene decided that it had had enough, and it could from now on live on marketing the growing agricultural produce of the county. The Texan trade was asked to move to other rail termini, and Wild Bill's contract ended.[17]

Here, then, is the historical development. The economic value of the plains was, to start with, marginal. At first they attracted primarily cow owners. However, the very act of pushing the cows into the plains raised their value, even if only a little. This rise in value justified a certain amount of Northern investment, such as the extension of the railroad or the settlement of new towns. This investment, in turn, raised the value of the land even further, so that an economy based purely on cows on the open range was no longer justified. The cow ecology would have to adjust to a much more competitive use of the land. And Texas, after all, did lose the war. The extermination of the bison made the Great Plains into a vacuum that Texas got sucked into—to get entangled in the web centered in Chicago.

Competitive use of the land was marked not only in the North-South confrontation of the saloon cities but also among the Texan herd itself. There were more and more cows now in an area devastated by the transition from the bison, and further reduced by the encroaching agriculture on its most fertile lands. To remain competitive, you had to secure the grass for yourself, and the cattle industry moved into buying and claiming land—fraudulently claiming homesteads or (especially in Texas itself, whose landholding system was not influenced by the Homestead Act) directly leasing vast estates from the state.[18] As the 1870s turned into the 1870s, control over cows had to be supplemented by control over the land itself. Where the cows would go was now something to be fixed. Farmers, as well as the railroad, needed to make sure cows would keep off their lands; cow owners needed to secure land to which their cows, and no others, could get access. All those cows, all this motion: it now somehow had to be delimited.

That, in itself, was nothing new. Ever since domestication, the control of animals on agricultural land had always been a complicated operation—especially considering that animals were in fact always forced to provide the main source of muscle power on the farm and as a result were part of agricultural production itself. Modern capitalism insisted on this discovery: that private ownership of the land could lead to intensive investment and thus to much-higher profits. The enclosed field, kept for the use of a single owner, was one of the hallmarks of capitalism in Britain and therefore also later in the United States.[19] Such fields could have been defined symbolically but more often were fenced by some combination of various materials easily available in northern Eu-

rope and in northeast America: stone, hedges, and—especially in rain-rich Atlantic America—wood.[20] With the expansion of agricultural land in America, the capital represented by fencing grew immensely; an often-quoted report by the Department of Agriculture in 1871 put the total value of fences at over $1.7 billion, and the annual cost at about $200 million, so that "for every dollar invested in live stock, another dollar is required for the construction of defenses to resist their attacks on farm production."[21] The purpose of such reports was to suggest a transformation in the use of fences: instead of keeping animals out of fields, they should be used to keep animals inside given boundaries, so that agricultural fields could make do with more symbolic fencing. Here was the trouble, then: such symbols would not do with animals, and no police force would buttress such a symbolic definition of space to make it effective to prevent animals' motion. They must be stopped by force; but where would the money come from to stop those herds of animals in this vast new land of the West?

The West had all those animals, and it did not have the means to stop them. The traditional materials of fencing were scarce, the most traditional of them—wood—nearly absent. Even the earth, a dusty, crumbling land—ideal for the grass—did not produce the right sort of stones. And while hedges could and were grown, they had their own limitations. As George Basalla, a historian of technology, notes for Osage oranges (the most prominent hedges of the American West in the 1860s and 1870s), "They were slow to develop, could not be moved easily, cast shadows on adjoining crops, usurped valuable growing space, and provided a shelter for weeds, vermin and insects."[22] In other words, hedges were inappropriate for the special colonization process going on in the American West, in which vast stretches of land were brought under control during a brief span of time, and the entire process was to be achieved with maximum flexibility and profit. This was colonization driven not by the life cycles of growing populations but by the expectations of capitalist investment. The three to four years it took an Osage orange to grow (as well as its element of waste) now became a major drawback. Four years, in the life of the plains, could be an eternity—the time it took, for instance, for Abilene to become a center for the cow industry and then to get out of that industry altogether. Geography was shifting daily. Something had to be found, quickly, to control the cows.

Fencing materials had to be imported, and the growing rail network—the essential infrastructure for the entire growth of agriculture in the West—transported those materials. Wood was shipped to the West in vast quantities; after all, American houses were built of wood in the West, just as they were in the East, and the railroad itself consumed timber.[23] But the vastness of the spaces involved made such shipping doubly unprofitable for wooden fences—both because of the vastness of spaces to be enclosed by such fences and because of the vastness of the space to be traversed by railroads put in place to carry such loads.

Thus a new technology for fencing was made a necessity—as stressed by the literature on the invention of barbed wire. But notice that the necessity was made by people, not given by nature. It did not derive from sheer geography—the presence of this space, the absence of those woods. It derived from the way in which America sprang upon the West, to enmesh it, almost in an instant, into its economy.

But let us return to the problem as it was perceived by individual Americans. They confronted animals; they were trying to control them. Such a task could be conceived as a kind of education: how to get an animal to do as you wish? This is essentially how the task was perceived in 1873 by Henry Rose, a farmer in the village of Waterman Station, Illinois. Trying to control a "breachy" cow, as he referred to her, he conceived of the following plan. He attached a wooden board, studded with sharp pieces of wire, right next to her head. Thus the cow, he reasoned, would be cured of her mischievous tendency to pass through loose fences. Now whenever she tried to squeeze herself through a limited space, pushing against barriers, she would cause herself considerable pain.[24] Of course, the idea of education through pain was familiar. Children at the time, after all, would regularly have their bare feet lashed with hickory sticks for failing to remember the multiplica-

tion table.[25] Children needed to learn arithmetic, and animals needed to learn fences. There are even specific precedents for Rose's experiment: for instance, we may compare it to the triangular yokes with which hogs were collared in seventeenth-century Massachusetts.[26] These yokes were intended to prevent hogs physically from crossing through fences, rather than to make pain ensue from such attempts; but essentially Rose's idea was an extension of the idea of the collar and similar attachments to the bodies of animals. The wooden board served as a tool for constant surveillance and punishment, even in the absence of the human.

After a while, it occurred to Rose that the fence itself could teach its own respect, serve as its own surveyor; instead of the sharp wire being attached to the cow's head, it could be attached to boards of wood on the fence itself. The experiment made, Rose was satisfied: the cow learned not to approach the fence. Other Americans made similar trials during the same period. Adrian Latta, for instance, attached sharp spikes to the bottom of his family's fences (he himself was only ten years old at the time, 1861) to prevent hogs from crossing underneath. He noted that "the hogs got through a few times after the barbs were put in. However, the barbs had the desired effect as the owner saw his hogs were getting terribly marked and kept them at home."[27] So instead of education, Latta's aim was sheer violence—aimed directly at hogs, indirectly at humans. If Latta's inspiration was perhaps nothing more than juvenile sadism, William D. Hunt took as an inspiration the venerable idea of the spur. This ancient invention consists of a roughly cut piece of metal that, thrust against the flesh of the animal, goads it to abrupt reaction. Hunt's patent, issued in 1867, positioned spur wheels on wire. The animal, pushing against the wire, would be wounded, though real injury would be prevented as the spur wheel turned under the animal's thrust. This, in retrospect, was a mistake: Hunt's spur wheels were, so to speak, too lenient, so that animals were not ultimately deterred by them. The same went for Michael Kelly's patent in 1868: cut nails thrust into wire. Once again, the nails would simply rotate on their wire when pushed against by the animals. Still,

Kelly was sufficiently concerned about the injury this might cause to animals that he called for tarred rope to be strung along the fence so that animals could see it in the dark and not get accidentally injured.[28] The peculiar experience of Henry Rose was meaningful: by starting from a corrective collar, he was prepared to the fundamental idea that, by causing pain, the fence could create the habit of its own avoidance. The genius of the new technology was that—once again—the cow's habits and skills were enlisted against her. Rose's fence acted not on the cow's skin alone but also on her memory and judgment, and these were ultimately used for her control. No need for the farmer to constrain his violence, then; make the cow feel the pain, and she will do the rest. Ultimately this was how hedges functioned—and the Osage orange, in particular, was protected by sharp, strong, long barbs, in retrospect highly suggestive of barbed wire.[29] Rose patented his idea and took it to be displayed in a farm exhibition in De Kalb, Illinois, near his hometown.

Notice that, with Rose's invention, iron barbs supplemented wood and did not replace it. However—as we have seen with some alternative inventions—others were already experimenting with materials. In the mid-nineteenth century, organic components gave way rapidly to their metallic counterparts. Iron production was exploding, and the material was searching for applications. In 1852, Samuel Fox invented the use of wire ribs for the frames of umbrellas—a huge British industry—replacing whalebone; the same was happening in the (generally similar) industry of corsets. Staying in the same domestic setting, we may take the production of strings for musical instruments; here, once again, wire became cheap enough around midcentury to begin to replace the guts of sheep. With the mass production of steel wire at midcentury, the piano began to be mass-produced as well—a major development in European culture. Umbrellas, corsets, and pianos were all important industries. Closer to home, though, for the interests of American farmers, was the production of rope, revolutionized in the 1840s as hemp began to be substituted by iron. Iron strings had tremendous strength, and most important, ma-

chines could be produced to automatically strand such strings into a braided rope.[30] This immediately suggests an idea: if the linear strings can be twisted together to construct stronger ropes, attached linearly, they can also be netted together, on a planar pattern—a fence made of wire. These, of course, are a familiar feature of our own contemporary landscape: fences made of woven lattices of wire, once again an invention of the 1840s and the 1850s. Butts and Johnson from Boston, for instance, advertised their "patent wire fencing" in 1856 "for enclosing railroads, canals, fields, cattle pastures, cemeteries, gardens, heneries, and for ornamental garden work, grape and rose trellises, etc."[31] Whatever were the real hopes of Butts and Johnson, such lattices did not fence in cattle: these structures are rather delicate and are made even more vulnerable by the contraction and expansion of iron under changes in temperature. Even determined humans can, with patience, run down such fences; they are no obstacle at all for herds of cows.[32]

It is here, finally, that barbed wire comes in. One of the visitors to the De Kalb fair, Joseph F. Glidden, formed the following idea: instead of attaching Rose's barbs to wood, they could be coiled around one of the strands in a double-stranded wire. The double-stranded structure, as well as the coil of the barb itself, would keep the barb in place. In short, unlike previous inventors, and emboldened by Rose's idea, Glidden decided to make the barb fixed so as to resist the cow in its approach and to inflict real pain. Further, Glidden's main technical idea—stranding two wires and a series of barbs between them—came from the experience of stranding wires together to form ropes, where the crucial fact was that machines already existed for the operation. No special new ingenuity was required: standard practices could be extended to achieve the mass production of barbed wire. And this is how barbed wire was born. In a sense, it was a natural idea, the confluence of all that went into the West: violence and the need to control space, iron, mass production. At any rate, a number of visitors to Rose's exhibition at De Kalb went away with the idea of attaching barbs to iron fences instead of wooden boards.

Glidden's original patent was quickly joined, apparently independently, by five other barbed wire patents, and most began production almost immediately.[33] Already in 1876, half the rights in one of the main patents were bought by Washburn and Moen, a Massachusetts-based iron and steel company, and in this way barbed wire reached the mainstream of manufacturing industry.[34] It was an extension of existing technologies, and so, although it had been invented on the prairie, it was soon taken up by the mass producers of steel and iron.

Washburn and Moen knew what they were doing. Barbed wire was an instant success. In the spring of 1875, the first commercial leaflet produced by the fledgling Glidden company claimed that the fence had been tested already by more than a thousand farmers—hardly a hyperbole, as the statistics available from later in the decade would indicate. Some of the selling points Glidden made were especially interesting:

It is the cheapest and most durable fence made.
It takes less posts than any other fence.
It can be put for _ the labor of any other fence.
Cattle, mule and horses will not rub against and break it down.
The wind has no effect upon it and prairie fires will not burn it up.
Stock will not jump over or crawl through it.

Two major claims emerge. First, the technology had the advantage of violence, so that it more effectively protected the space it enclosed. Indeed, not only was it a kind of fence that protected the inside it surrounded—but the fence protected itself as well. Second, the technology had the advantage of iron. The material was more resistant to natural forces. Combining the two, the power of violence and the power of iron, led to the major advantage of the technology at that stage. Lighter materials were required now to construct a fence, hence less labor, hence ultimately the fence's competitive pricing.[35]

Here is how Washburn and Moen were to state the case in 1880, in one of their early pamphlets. Taking forty rods of three-row fences as the unit of comparison (about one hundred meters of fenced length), we have the following:

WOODEN BOARD FENCE		BARBED WIRE FENCE	
1,000 feet pine fencing	$15.00	136 lbs barbed wire	$14.96
80 posts	$16.00	40 posts	$ 8.00
15 lbs nails	$ .60	2 lbs staples	$ .20
Labor	$ 2.50	Labor	$ .50
Total	$34.10	Total	$23.66[36]

The beauty of this pricing scheme is obvious: the main component, the one on which Washburn and Moen make a profit—barbed wire itself—is priced artificially high, just below the price of the main alternative piece of hardware, pine. The entire barbed wire fence is made competitive only because barbed wire, especially owing to its lightness, is cheaper to erect: fewer posts and nails are required, and much less labor. Notice, however, that posts *are* required—and were usually made of wood. This is an important aspect concerning the growth of barbed wire: it did *not* replace wood. That is, barbed wire did not at all result in a *reduction* of the importation of wood to the West. It did not, after all, replace existing wooden fences: instead, barbed wire fences were erected where no fences had been erected before (and none, probably, would have been erected otherwise). Thus barbed wire actually led to a *growth* in demand for timber. As the West was becoming capital intensive, the North was being deforested.[37] Barbed wire represents therefore not the replacement of wood by iron but rather a more effective combination of the two. It uses wood for its capability to sustain weight, and iron for its capability to take on precise, strong forms. In this, barbed wire resembles the two other typical technologies of the period, the railroad and the telegraph line, all consisting of repeated bases of wood, set perpendicularly to support long lines of custom-made metals. (The telegraph, another huge wire-based industry, used copper rather than iron.) Short posts of wood, repeated at regular intervals, supply these objects with a solid hold on the surface of the Earth; metal lines, attached continuously, make them stretch without a hitch to an indefinite length. In combination, such objects can accomplish a task, defined along immensely long lines, and in this way they reshape space—railroads and telegraph lines by connecting distant points, and barbed wire by defining lines of limit. This is the material context in which the growth of barbed wire should be placed.

The spread of such lines determines the transformation of space. From 9,000 miles to 30,000 miles of tracks: this was the growth of the American railroad during the 1850s, the period during which northeastern America was forever reshaped by train.[38] By 1880—a mere six years after its first patents—something like 50,000 miles of length were fenced by barbed wire.[39] We are therefore justified in comparing the revolution of barbed wire with the revolution of railroads; both transformed space almost instantaneously. The difference lies in the intended species—and is also the difference between lines as connectors and lines as dividers. While the purpose of trains was to make motion *possible*, for *humans* (as well as for their commodities), the purpose of barbed wire was to *prevent* motion, for *animals*.

The key to the entire success of this technology was, of course, its ability to stop cows. We have looked at how manufacturers priced (or at least attempted to price) their barbed wire, but whereas prices could be fixed on paper, animals had to be stopped in the real world, and humans had to be shown that. This was indeed the marketing strategy employed by the distributors of barbed wire. Let us look at an event that acquired an emblematic significance in the literature on barbed wire: the exhibition in San Antonio in 1876. Three years earlier, the farmer Henry Rose had displayed the fruit of his farm experiment on his cow. Now the salesman John Gates offered a far more striking spectacle. (Gates, at this point, was a mere agent for barbed wire manufacturers, but he was destined—as we will see in the next section—to become an international iron magnate.) Part of the central plaza next to San Fernando Cathedral was surrounded by barbed wire fences, and dozens of fierce-looking longhorn bulls were packed into this space. Here is the view to have greeted you, stepping out of the cathedral: the animals deliberately frightened and provoked; they charge against the fence and are repulsed by the sheer pain of sharp metal tearing flesh; they are wounded, and their wounds exacerbate their rage; further charges, further pain, and instinctive withdrawal; finally, resignation. The spectacle was indeed symbolic: it showed how, without the slightest touch by humans, the fiercest, and at that point least domesticated, bovine animals could be made to respect a definition of boundary: the bulls learned to avoid the fence. Barbed wire could succeed as a tool of surveillance and education.

It became clear that cheap, flexible, and effective means were available to control the movement of cows, even without the need for human intervention. Sales skyrocketed—from 5 tons in 1874, the first year of production, to 300 tons in the following year, surpassing 10,000 tons by 1878 and 100,000 tons by 1883.[40] Of course, the artificial prices that the manufacturers tried to keep did not hold. Even in 1880, actual prices paid were about half those mentioned in Washburn and Moen's advertisement quoted earlier.[41] By 1885 the price was halved again, and by 1897 it was more than halved again. At that time, the original patents began to expire, driving prices down even further. This made the invention of new patents an attractive business, and new technological advances made the technology more economic and effective. Most important, it was realized that steel, while more expensive to produce per ton, could resist animal power with much lighter strands, so that overall the price per mile would be lower with steel than with iron.[42] Steel barbed wire became more and more common during the late years of the century. (Its greater power would be significant during the next century, when barbed wire was to meet humans rather than cows.)

Instead of being a prohibitive element of cost, fences now became a cheap, labor-efficient resource, and so fencing could be extended not only in space but also in its intended uses. It is probably true that barbed wire was invented in Illinois with the farmer in mind, to protect his fields from animals; but almost immediately, barbed wire was used by owners of cows. Just as they were struggling with the diminishing resources of grass and water, they were now saved by the growing resources of iron and capital. The central fact was that as the economy gradually made its upturn from the financial panic of 1873, capital began to pour into the West, driving more and more cattle there, attempting to acquire more and more land. From 1876 onward, more than 200,000 cattle were moved annually north from Texas to form the basis of new ranches.[43] Hundreds of companies formed, mainly on the Atlantic seaboard and in Great Britain, attracted to what was perceived as a bonanza.[44] The usual logic of concentration of capital applied: as pointed out in a government report at the peak of this process, "generally it is found that the average cost per head of the management of large herds is much less than that of small herds. The tendency in the range cattle business of late years has therefore been toward a reduction in the number of herds, and generally toward the consolidation of the business in the hands of individuals, corporations, and associations."[45] Larger herds require larger space, and the larger a space is, the smaller the ratio of unit of perimeter to unit of area. This, then, was a further crucial element of economy: fencing became cheaper, paradoxically, on the immense units of space used by large herds. It also became more necessary as the same spiral of overuse continued to force fiercer competition for resources. The result was that lines of fences were set to define territories on which companies grazed their cows—whether those companies had legal title to that land or not. In 1885 it was reported that almost 4.5 million acres had been illegally fenced in this way.[46] Illegal as well as legal fencing led to wire cutting, typically as owners of smaller herds entered spaces controlled by owners of larger herds, to use their grass and water and, frequently, to steal their cows. Warfare surrounding wire began here: from Texas to Montana, big firms owned by Atlantic capitalists fought against adventurers who thought they could still make big money on violence alone. But this was not the Texas of the 1830s or even of the 1860s. Gangs of hired guns were employed by the big firms and provided with lists of small-time rangers to be killed.[47] The big-ranch business was busy, in short, driving out the range business (and the small-ranch business), much as the cow business had earlier driven out the bison business. The comparison is meaningful: in a sense, cows were driven off the land just as surely as the bison had been. The only difference was that whereas the bison were killed, the cows were imprisoned, in an Archipelago Ranch, so to speak, strewn across the plains.

To illegally fenced land, one should add several million acres of legally fenced grazing lands (especially in Texas, where the legislation was more favorable to large landholding). A notable example was the XIT Ranch in the Texas Panhandle, named after the ten (X) counties through which the ranch extended; by 1885, it had 50,000 cattle fenced inside 476,000 acres. This may be compared with the size of the territory owned in the West through the Homestead Act by 1884—just over 16 million acres.[48] In short, then, about a decade after the introduction of barbed wire, it was already used to surround cows as much as, and even more than, it was used to surround farmlands. All of this, it should be stressed, had never been envisioned by Rose, who had thought in terms of the age-old confrontation between a single farmer and a single animal. The mass scale of it all came out of the West itself and, in this way, re-created the tool beyond the imagination of its inventors. It could now be used to redefine space itself.

A striking example of the new manipulation of space made possible by this new technology was drift fences. Faced with a bad turn in the climate, what cows remained on the range would instinctively turn south—to compete there for diminishing resources. Texans (and Oklahomans) therefore looked for ways to prevent the motion of cows from north to south, in particular since the same kinds of climatic conditions that made such prevention desirable also made it difficult for humans to stay on the open range and to stop the animals in person. Beginning in 1881 or 1882, therefore, a new type of fence was gradually built: a long line of fortification against the North, as it were—built by many individual landowners, with little coordination, but effectively constructing, ultimately, a barrier across the entire Texan Panhandle. Those fences worked; cows, even those that were not yet fenced in, could be relied upon to stay in the North and not to compete for southern grazing land. No cow was now free from having its motion limited by barbed wire, and the basic geography of the land was now redefined. The climax of this development came in the severe winters of 1885–1886 and 1886–1887. Heavy blizzards drove cows in many tens of thousands southward, but just like the bulls of San Antonio, they could not pass through the barbed wire; weakened by the storm, they now were wounded and frustrated. Even in their concentrations of hundreds upon hundreds, they could neither break through the barrier or generate enough warmth between them to survive. Trapped like that, they died, perhaps as many as two-thirds of the cows in the open field, dying of starvation and of cold. Not all victims, of course, reached as far south as the drift fences, but those that did—the ones who had the most chance to survive—were perhaps those that suffered the most.[49] Such images—piles of the dead, huddled together, desperately crushed against barbed fences—are eerily reminiscent of twentieth-century images (I will return to such historical continuities later in the book).

In short, barbed wire was a success. It could stop animals, no matter how many, no matter how desperate. At first it was not clear that the static violence of barbs alone could make wires sufficiently effective, so manufacturers erred on the side of caution, supplying sharp and large barbs. These barbs met semiferal animals, accustomed to free roaming. Crashing against the wire—as they did at Gates's display at San Antonio—cattle would inevitably get seriously gouged. Open wounds ensued, which in the warm, humid summers easily lead to screwworm infestation. The screwworm fly was an endemic part of the southern plains cow economy. Its life depended on the wounds of large, warm-blooded animals. In those wounds the female would lay its eggs. When they emerged, the worms could literally eat the animal alive. All of this was extremely disagreeable to the cow owners: "A particularly disgusting and sickening job was when cows or calves got screwworms in their mouth or gums. . . . [This some-

times happened when] [t]he cow or calf—if they could reach the wound—would try to lick the worms out of the lesion. . . . You couldn't use any medicine—just remove the worms and hope you got them all. It sure wasn't a job for anyone with a 'queasy' stomach."[50] The owner's concern was not just aesthetic. Besides the loss of value owing to the death or severe illness of cows, screwworms were a severe drag on the cow economy in two ways: they demanded skilled labor, and (since birth invariably led to wounds) they made it undesirable to allow birth during the summer, thus curtailing the natural growth in the number of marketable cows.[51] The plains were rapidly moving away from the simple nature-turned-into-profit of the mavericking days. Barbed wire created the conditions for a new type of cattle industry; simultaneously, it was a constant source of loss to it. Thus we should not be surprised that, among some farmers, barbed wire was unpopular. Farmers in the late 1870 were surrounded by barbs they had not asked for, causing damages they could not control. Now, an interesting feature of barbed wire is its symmetry. While it is possible in principle to have barbs arranged so that they point in just one direction, it is far simpler to have them double pointed, so that the wire can be made "blindly," without figuring how the barbs precisely fit. In other words, the topology does not distinguish "inside" from "outside"—violence is projected in both ways. In a very real way, barbed wire is contagious: by enclosing a space, it is thereby automatically present in all areas bordering on that space. Imagine that you are a farmer, used to controlling your animals without barbed wire, now finding yourself adjacent to it. This could happen anywhere, especially since almost from the beginning, the railroad had used barbed wire to prevent animals from straying onto the tracks and causing damage to the trains.[52] Take, for instance, Mr. Palmer, a representative of Jericho to Vermont's General Assembly, who in 1880 drafted a bill to limit the use of barbed wire. He focused on the injuries caused to horse and cattle by railroad barbed fences and concluded that "the public sentiment of the community was against its use in these three cases: along highways; between adjoining landowners without mutual consent; and between railroads and pastures without the consent of the farmer."⁵³ Bills to prohibit the use of barbed wire were put forth in several states, always defeated in the West but, for brief periods, made law in some eastern states. Mr. Palmer actually succeeded (as did some of his colleagues in Connecticut and Maine), but by the end of the 1880s, no state made barbed wire illegal any longer; the plains had reached the North.

Some compromise had to be made between the conflicting interests. In fact, following the first years of violent encounters between animal and iron, a new relationship was gradually established. To some extent, barbs became less sharp (less "vicious," to use the technical term). It is instructive to compare the declared objectives of barbed wire patents between their introduction and their later accommodation to farmers' needs. In 1876, Parker Winemann of Illinois boasted of his barbs that they "will be sure to penetrate the skin and give pain",⁵⁴ five years later (i.e., immediately after barbed wire began to be politically contested), Joseph H. Connelly of Pennsylvania stated that his particular invention "will resist force and turn stock without entangling or otherwise injuring them."⁵⁵ One such invention is typically praised in a Washburn and Moen leaflet: "the barbs are short and lance shaped, so that there is NO DANGER OF INJURY TO STOCK. . . . They will prick and scratch but NEVER TEAR THE SKIN. . . . It is well known that the sensation of pain is at the surface of the skin, hence the smart or prick . . . is all that is required. NO WOUNDS ARE MADE, consequently NO LOSS OF CATTLE in the southwest from putrefying sores, in which flies deposit their eggs."⁵⁶

As humans learned more about animal pain, animals learned more about human violence. Animals learned to avoid barbed wire, and sometimes they were deliberately taught. A commercial leaflet encouraging the use of barbed wire advises the farmer "to lead [young horses] to the fence and let them prick their noses by contact with it . . . they will let it thoroughly alone thereafter."⁵⁷ This expresses the special interest humans always had in the physical fitness of horse, whereas cows were generally expected to pick up such knowledge through sheer experience. Their knowledge was apparently transmitted between generations, by the experience, for example, of calves following their mothers (one should remember that the cow economy of the nineteenth century still did allow calves to grow up following their mothers). To close this circle of mutual knowledge, finally, the stage was reached where manufacturers, exploiting the knowledge gained by animals, produced more conspicuous barbs, now functioning not only as instruments of direct violence but also as a more indirect instrument of intimidation (in the technical language, barbs became more "obvious"). This transition was essentially complete by the end of the 1880s, when the success of barbed wire as a tool for the education of cows can be considered complete. Simultaneously, more and more cows were fenced in, rather than fenced out, partly because of the general trend to establish landholding, and partly to "protect" the cows. The topology was now inverted, just as it had previously been for the Indians. Instead of fences preventing the motion of cows from outside a closed line to its inside (protecting the property of farmers), fences now prevented the motion of cows from inside a closed line to its outside (imprisoning cows inside ranches). Fenced inside, cows could be taken care of in the case of screwworm infestation, and winter catastrophes could not return in such harshness—so that, to a certain extent, cows were fenced in to protect them from fences.⁵⁸ Cows and plains were transformed, so that barbed wire became both natural and necessary.

With the gained perspective of nearly a decade of barbed wire use, Washburn and Moen adopted an almost historiographical tone in their commercial leaflet of 1883, already quoted briefly. The analysis offered is especially sharp and acute, and it is worthy of lengthy quotation as a summary of the early development of barbed wire:

> The fence of plain wire was far from satisfactory. . . . It had no terrors for trespassing animals. . . . [S]omething else seemed to be needed to realize the perfect fence, and this came in its own time, in Barb Fence.

Barb wire was invented by a farmer, to meet farmer needs [it should be understood that Washburn and Moen, having won the ranch business, were now busy introducing barbed wire into cultivated areas, hence the stress on the "farmer needs"], in 1873, at first a crude working out of the parent idea; the making of fence wire repellent by borrowing from nature the principle of the sharp pricking thorn, thus appealing to the sense of pain and danger that resides in the skin of the farm animal.

The principle which was first sharply challenged as cruel, has, on the contrary, been found to be humane, for these accidents of the old style were common [a litany of non-barbed-wire fence complaints follows]. . . . [T]he accidents from barb wire have been mainly of a trivial character [in a brilliant rhetorical move, following the description of accidents involving traditional fences, the author is now able to refer to wounds caused by barbed wire as "accidents"—even though, of course, the causing of wounds is the essential function of barbed wire], which, in such cases, have been warnings, salutary in their effect, and have educated the beasts in the new law of respecting fences.

This, then, was the basic function of barbed wire: a form of education—of manipulating animals—through violence. From a broader perspective, we may conceive the manipulation, or transformation, as follows. As mentioned earlier, cows brought to America by the Spanish multiplied immensely on the loose, so that a new breed of semiferal cows was created on the margins of Spanish settlement, especially in Texas. This breed was gradually redomesticated by Anglo-Americans, until faced with the need to use those animals in the conquest of the West, the process of redomestication had to proceed much more quickly. Barbed wire served to retame, by a shock, an entire breed, partly through its immediate impact, and partly through its indirect biological effects. Fenced cows could be bred in a more controlled way. Ranches, defined in space, became also defined in stock (and gradually, as a rule, limited in numbers).[59] Breeding generally took the form of the introduction of bulls from eastern states—a docile and fat breed.[60] Backed by eastern investors, eastern owners came to hold more and more of the ranches, so that eastern cow-handling practices gradually became dominant.[61] Eastern capital, eastern iron, eastern semen: all were pouring into the West to turn it into a new, artificial land for the use of the East.

All of this, let us remember, was based on a simple fact about Texan cows—indeed, about most animals. Therein lies our misfortune: our skins, just a little beneath the surface, are endowed with special nerves activated by pressure rising above rather low thresholds. You can use those nerves against us. By cutting through the boundary of our skins, you can act to protect the boundaries of your property, your prison, your border. Iron, for instance, is a useful tool. It is harder than flesh; pressed against it, iron will first push the flesh inward and then (particularly if the iron's surface, like that of a barb, is sharp) cut through the skin to impact on the nerves. The nerves send a report to the brain, and there the report undergoes some process—we do not know quite what—that leads to something else—we cannot explain quite what. This is what we call pain, and apparently it is something truly universal, cutting across species, places, and times. A useful tool of globalization, then.

3. HOW TO FENCE THE WORLD

Joseph Glidden was not so sanguine. A cautious capitalist, worried that his invention might fail through bad marketing, he kept sending anxious letters to his agents. On September 15, 1875, he was admonishing Sanborn (who had shown signs of straying from the plains marketing strategy): "we do not expect the wire to be much in demand where farmers can build brush and pole fences out of the growth on their own land and think the time spent in canvassing such territory very nearly lost even if some sales are made."[62] Sanborn should stick to his domain, the plains of Texas. Glidden's perspective, writing from De Kalb, Illinois, was defined by the plains, but as soon as production moved to Massachusetts, the perspective widened dramatically. New England had an experience of world commerce; barbed wire now was to join this global trade. Already in 1877, Ferdinand Louis Sarmiento was working in the South American continent as an agent for Washburn and Moen, busily seeking outlets for barbed wire.[63] In December of that year, for instance, he managed a public relations coup: Carl Glash, director of the Imperial Botanical Gardens in Rio de Janeiro, issued an endorsement of a barb fence erected there that he found "exceedingly complete and useful." This fence, probably one of the first to be erected anywhere outside the United States, "was erected to enclose a collection of rare water and wild fowls and animals brought home by H.I. Majesty Dom Pedro II from his late North American and European trips." It did not take long for barbed wire to reach beyond the botanical gardens. Essentially, barbed wire was used, at first, wherever conditions approached those of western America. The most obvious parallel was Argentina, where the huge plains of the Pampas were held by semiferal and feral animals, as well as by Indians. The Argentinean Indians were totally destroyed by 1879, and throughout the 1880s, fenced cattle—and wheat fields—covered the Pampas.[64] Parallel historical circumstances made the Pampas similar to the Great Plains. This similarity could now be exploited by American producers, who moved in to supply the Pampas with the technologies of the Great Plains. Soon barbed wire defined the Pampas just as it did the plains. It was estimated that by 1907, barbed wire in Argentina was already sufficient to surround the perimeter of the republic 140 times.[65] Everywhere, Washburn and Moen were aggressively seeking the prairie lands of the world, sending out powers of attorney in 1880 to diverse places such as Tasmania and the other Australian provinces, New Zealand, Cuba, Ceylon, and Russia.[66] In their correspondence, they repeatedly stressed how much further growth could be expected as a function of the global area covered by plains throughout the territories to which they had extended their patents. In 1884, when Washburn and Moen were working on a global arrangement with Felten and Guilleaume, from Mulheim, Germany, they claimed that they had already obtained patents "covering [Australia], New Zealand, India, Italy, Sweden, Austria and Denmark, representing a territory in the prairie countries only of those named . . . of 5,470,952 square miles in which no barbed wire can be sold without direct infringement . . . a territory compared with the territory of the United States as two is to one, or ensuring as soon as barbed wire is properly introduced in said countries at least 300,000 tons of sales per annum." (It would seem that Washburn and Moen had, somewhat disingenuously, measured Siberia as an equivalent of more fertile prairie land, ultimately revealing an almost prophetic vision for the future of barbed wire.)[67] In the end, Felten and Guilleaume obtained the following arrangement, which I will spell out so as to give a sense of the massive amounts involved—and how these were to be calculated across the globe. Felten and Guilleaume were to produce in the United States up to 1,000 tons per year, to be sold outside North America. They were also to produce, without limits, anywhere in Europe (or in other countries where they owned rights), paying Washburn and Moen two dollars for every ton sold in Britain and one dollar for every ton sold elsewhere, except for the amounts of 250,000 tons in both Germany and France, which could be produced and sold for domestic consumption free of charge to Washburn and Moen. Try to concentrate on that figure—a quarter million tons, almost as many miles—sold in Germany and France alone! This arrangement was continued in 1889, with the Germany/France clause changed to allow the sale of up to 1,000 tons anywhere in Europe, for domestic consumption, free of charge.[68] Yet at this stage Felten and Guilleaume were already Washburn and Moen's minor global partners. In 1891, Felten and Guilleaume's royalties paid to Washburn and Moen came to a little less than half those paid from Johnson and Nephew in Britain.[69] Johnson and Nephew's factories at Manchester and Ambergate, with about a thousand workers,[70] were among the largest producers of barbed wire at the time—American production, as we will soon see, was rather more dispersed during that period. British producers were, however, rightly concerned: German products were now dominant in Europe, while America had become a net exporter of iron products rather than a market for them. The huge production of Johnson and Nephew was thus based on export to the British Empire, with important consequences for colonial war—as we will see in the next chapter.

An important area for the introduction of barbed wire was Australia and New Zealand.[71] These continental areas were now opened for settlement, calling for some of the most radical ecological transformations Europeans effected anywhere in the world. Barbed wire reached Australia at a stage comparable not to the American cow but to the American bison: where human domination over other animals calls for extermination. Barbed wire's violence could easily be extended to extermination as well—simply by fencing water sources.[72] As for control over domesticated animals, the Australian tendency, at first, was to rely on plain wire. This was a sheep economy, based (as is usually the case with geographically marginal animal economies) not on the export of flesh but on that of other, less perishable animal parts—in this case, wool. Already in the 1860s farmers realized that wire fencing could be cheap and effective against such animals, and sheep with wire alone, barbs being introduced only gradually, later in the century, when the price of barbed wire was no longer higher than that of plain wire (in the twentieth century, of course, it was all barbed wire). Now that sheep could be perfectly controlled without an investment in the unreliable labor of shepherds, their numbers skyrocketed. From 6 million sheep in 1861, New South Wales had 57 million by 1894, now clearly the dominant animal on the land, with wire fencing the dominant land feature.[73]

South Africa—as usual—was even a more special case. Unlike other major areas of European colonization, it already had established agricultural practices, all based on animals. Black agriculture used the ox as the major source of muscle power and the cow as the major source of food. The Boers—descendants of seventeenth-century Dutch settlers—had already adapted European agriculture to South Africa. They gradually reverted to a form of low-

capital agriculture with strong dependence on pastoralism and hunting. All of this was now coming more and more under the power of the British Empire; Boer resistance to it would lead to important consequences for the history of barbed wire (to be seen in the next chapter). But even before formal British domination, Johnson and Nephew were taking over the land with their barbed wire. The land was ready for the change. In 1886, mines were discovered in the Witwatersrand—legendary treasures of gold. Almost instantly came the railroads, new immigrants, new cities. In the countryside, it was now more profitable to produce wheat for the urban market. The many governments of South Africa supported the new intensive agriculture; in 1890, for instance, the Orange Free State—a Boer government—made fencing a legal obligation. Relations of humans and animals were quickly changing. In 1892 the *Friend of the Free State*, an Orange journal, painted a vivid picture of the new methods of hunting game: "The modus operandi is to drive all the game against a farmer's fence and then shoot them down, regardless of course of the cost of the fence." These hunters were very inconsiderate, the journal suggested, but there was some hope in the future: "When fencing is more general, however . . . [they will have] to give up their favourite pastimes."[74] Soon an animal apartheid took shape: on the one side Boer and British cows, fenced in, on the other side wild animals as well as cows belonging to blacks, fenced out of the best lands. Barbed wire was a tool in the white landgrab, with the blacks removed to marginal, unfenced land. This could still feed all animals in good years, but when the rains failed, disaster would strike. No one can tell the exact impact on wild animals, but the impact on the black economy was obvious. By the century's end, more and more blacks were forced to become farmhands on white farms, a major step toward white ascendance. A new way of controlling the land, designed to make more efficient use of it, transformed the relations not only between humans and animals but also between different human groups—distinguished by their different access to the new technologies of control over space.[75]

Everywhere, the world system was building up its stocks of

barbed wire. The center was casting its net wider. Even American producers came to rely more and more on export. The dominant American producer from 1899 onward, American Steel and Wire Company, produced 34 percent for export in the first eight years of its activity (1899–1906), but 44 percent in the following eight years (1907–1914). Taking into account foreign production, it is likely that at the end of the nineteenth century, the point had already been reached where more barbed wire was installed outside the United States than within its borders. But still, the absolute importance of the American market should not be lost from sight: with more than 100,000 tons consumed annually, the United States was, throughout, the mainstay of demand for barbed wire.[76] It remained a leading net exporter well into the twentieth century. As late as 1932, barbed wire imports into the United States did not exceed more than 20,000 tons. About half of these came from Germany.[77] The very existence of barbed wire export from Germany into America was in fact significant: the Old World would not allow the New World to monopolize barbed wire.[78]

Even Europe, a growing barbed wire producer, was itself grudgingly becoming a consumer for its own domestic consumption (and not just that of the colonies). In the decades following Glidden's original cautious estimate, barbed wire returned to reshape old, established agriculture. J. Bucknall Smith, an engineer, sounded a note of alarm in 1891, writing from the perspective of British wire production. "Our American friends may run locomotives and trains through their public highways [but should we?]. . . Similarly, although barb-wire fencing is admirably adapted to the protection of landed property, and for enclosing live stock, in a large portion of the States or our colonies, &c., nevertheless we should scarcely be pleased to see it applied to our parks or promiscuously along our public roads." Smith had a deep insight into the historical process around him. The world was diverging—centers of polite society, where violence was now viewed with unprecedented disgust,[79] and, away from them, areas of economic expansion where unprecedented violence and power provided the profits to sustain polite society itself.

There is a paradox about the modern reshaping of space. Capitalism is based on spatial division of labor, assigning entire domains to a specific kind of production that cannot survive without interacting with the world economy as a whole. Thus it leads simultaneously to two opposing processes: as parts of the world become mutually dependent, they also diverge from each other. Barbed wire, contributing to the integration of the animal industries of the world with the world's urban centers, also formed part of the growing divergence between urban and rural. This had two aspects. First, the rural world was being unified; across the globe, different rural economies became part of the same system (as, of course, was happening at the same time to the cities themselves). Second, the rural world was, as a whole, pushed out of sight of the urban world, creating a major cultural divide.

The globalization of the rural world was keenly felt on the Great Plains themselves: they were now part of a world system, based on the urban centers of the northern Atlantic. This world system was not merely financial but also biological. I have mentioned the growing domination of the cow industry in the West by eastern breeds. This gave rise to concerns. It should be understood that historically, Texas was very different from the truly intensive cow districts of the world. In places such as, say, England, cows lived next to urban centers so that their milk could be transported and consumed by city dwellers. Thus a much more dense population of cows could be profitable. With its greater motion, a commodity in a world connected by rail and steamboat—this Old World of cows was also susceptible to new outbreaks of disease. A major epidemic of rinderpest in 1865 shook everyone in the cow economy. Shivering, frothing at the mouth, refusing all food, cows died in the millions—sometimes as many as half the herd.[80] The epidemic began to be monitored in Britain, quickly traveled on to the Continent, then to New England.[81] Texas, however—its cows herded much less intensively, and, in this period, almost a cow world unto itself—was spared. As Swabe has shown, the 1865 rinderpest epidemic was a major event leading to the new veterinary regime of the 1870s. In Europe, an attempt was made to control the motions of cows on the basis of science, the old practice of the quarantine applied with great force.⁸² This, then, was the background for the alarm of the United States Treasury Cattle Commission, expressed in Chicago on August 23, 1881: "That a very large proportion of our country has, up to this time, remained exempt from [rinderpest] is owing chiefly to the fact that the current of our cattle traffic has heretofore been mainly from the west toward the seaboard; but the business of purchasing calves from the eastern dairy districts and scattering them throughout the western states and territories, which has, within a year or two past, assumed such mammoth proportions, has augmented the danger . . . tenfold."⁸³ It should be noted that the commission had no regulatory power, and anyway, the commissioners had missed the point. What had saved the Texas cows was the isolation of the plains, but their very economic value now ended that isolation. Soon they would have (on top of the screwworm) all that cow flesh is heir to: rinderpest, anthrax, tuberculosis, foot-and-mouth disease—and many other diseases, filling the journals of veterinary science, for decades to come, with useless medications.

All of this was being segregated from the urban sight. A crucial development was the invention, in the 1870s, of refrigerated train cars. Refrigeration along a distance, for the first time, made it possible to build a spatial separation between the killing of animals and the life of humans. It should be understood how Sisyphean a task it is to kill an animal. You stop its heart beating, and still, billions of organisms go on thriving inside. You think you have gained full mastery over the animal by slitting her throat, but all you have done is to start a new battle, now for domination over the dead body. So now the dead animal has to be boiled, frozen, inundated with minerals—everything to kill the microorganisms. (More recently, radioactive exposure has been added to the arsenal in this fight.)⁸⁴ Ultimately this is a losing battle, and the longer you take between killing the animal and consuming it, the more likely it is there will be nothing left for human consumption at all. This is very unfortunate for the manufacturers, for, as I have

mentioned already, profits are always proportionate to distance. Historically, the animal industry could produce such distance-based profits only by severely limiting itself. Hides and other tissues of the body (such as horn), already semidead, can be used as something more akin to mineral resources. Hence the original killing of the bison. But this leaves out most of the animal's body. Heavily salting the animal is another solution, but this gives up the most lucrative business of the more upscale, raw flesh. That is the sadness of it, you see: people *like* the taste of blood. So you can, as an alternative, let the cow grow in a faraway place (to profit from cheap land, resources, and labor) and then transport it, still alive, somewhere near a center of consumption, to be killed there. But this implies investment in slaughterhouses on prime real estate (in the American context of the 1870s, this meant property in New York). There is also the wasteful need, already mentioned, to revive animals somewhat after transportation with expensive feeding near urban centers. No: a way had to be found to make the dead flesh of animals a raw commodity and to make it participate in the new network of transportation.

This network itself provided the solution, and once again, the American West led the way. Chicago became the meeting point for two commodities: cows from the plains, and ice from northern lakes and rivers. Cows would now be killed in Chicago and transported to the East. This invention, evolving simultaneously with barbed wire and reaching perfection in the late 1870s, ensued in a new, macabre railroad car architecture. The passengers were dead carcasses, closely packed together as they dangled from the center of the car like tuxedos on a huge coatrack. From both ends of the car, they were guarded by boxes of ice and brine; a ventilation apparatus blew frozen air into the compartment. These new cars created a buffer zone between a polite urban world, where animals were seen more and more as nothing more than meat, and a violent rural world, where the killing of animals became more and more profitable.⁸⁵ Now that the American killing of cows was nearly all concentrated in Chicago, killing and processing could benefit from the concentration of capital. Huge factories were built, based on what came to be known as the "disassembly line." Living animals were transformed into so many products, for although iron replaced so many organic resources, such resources were never discarded. The animal that was not eaten had to be used. Some of it went into products used by humans—buttons, for instance, made of bone. These would ultimately be replaced by synthetic products such as plastic; more significant in the long run was, so to speak, the recycling of the animal, that is, using its carcass for such purposes as fertilizer and animal feed.[86]

At the two ends of the carcass trains, worlds were disengaging. Agriculture was intensifying, and the lives of plants and animals were now spent in worlds created by the agronomists' manuals, far from the imagination of the city. This, in turn, drew away from the animal. Muscle power was reduced in value, especially as the millennia-long tying of the horse to the wagon was giving way, by century's end, to the forces of steam and electricity. The streets saw fewer horses. They also saw fewer animals brought to be killed (once a typical urban sight). The town butcher became a dealer of packed meat, and killing was relegated to faraway meatpacking factories.

Which is all to say that control over space was ever more perfected. Raise in Texas; kill in Chicago; eat in New York. Or raise on the Pampas; kill in Montevideo; eat in London. The northern Atlantic had now truly dominated the spaces of the plains. Let us not forget the true order of causes. To begin with, the inhabitants of the northern Atlantic shared a dietary heritage that prized the flesh of bovines.[87] (In the next chapter, I will explain why this was the case.) In the early nineteenth century, Americans liked to eat beef à la mode, which was ground cow flesh, incised and stuffed with bread crumbs, spices, and butter (of course also a product of the milk of cows). Most common was the fried steak, often served—in affluent northern cities—for every meal, breakfast included.[88] This breakfast steak was the ultimate cause of all we have seen so far. In America, it was not the West that shaped the eastern diet; it was the eastern diet that shaped the West. The same elsewhere: that so much of the globe was now given over to growing cows was an expression of how much the world was governed from Boston, New York, London, and Berlin. No rice eaters, there.

No sentimentalists, either. Americans embraced the violence of barbed wire, just as they embraced the violence of competition to which it gave rise. In the dynamic years of the American steel and iron industry, barbed wire was crucial—the spur that pushed the industry in its most significant development. We have seen how half the rights in Glidden's patent were bought already in 1876 by the Massachusetts iron producers Washburn and Moen. But with the technology being so simple and lucrative, it was easily pirated. In the ten years after its invention, at least 114 companies were formed for the purpose of barbed wire production. All they needed to do was to buy a wire-stranding machine, invest in plain wire from some of the big companies in the East, and hope for good marketing in the West. Unable to drive the competitors out of the market, Washburn and Moen finally succeeded, in 1880, in making them all into licensed producers for Massachusetts itself. There were still many producers, but they had agreed to set production quotas and pay royalties to the holders of the patent. During the following decade, however, competition between the many producers did drive the price of barbed wire down—and so drive many producers out of business altogether. Price of plain wire was falling much more slowly, and margins in the industry reached critical levels. The patent expired in 1891. To stay profitable now, the few remaining barbed wire producers had an enormous incentive to reach arrangements with plain wire producers. We have met John W. Gates already, in 1876 on the plaza in San Antonio, a barbed wire salesman goading bulls into being gouged by barbed wire. He went up in the world. Now he was gouging the industry, in 1892 creating the Consolidated Steel and Wire Company, a $4 million holding company with two wire mills and three barbed wire manufacturing concerns. The years after 1893 were a period of depression. Now it was the wire mills themselves that felt the squeeze. For the wire industry to survive, it had to consolidate so as to cut its costs. Gates easily convinced more and more wire mills to consolidate with him. In 1897, finally, he approached J. P. Morgan in person. Morgan, one of the wealthiest American capitalists, was asked to underwrite a huge conglomerate encompassing most of the American wire business. Morgan nearly agreed the following year, when a war with Spain broke out over the island of Cuba (I will return to this war later). Not the time to invest in wire, Morgan considered. Gambling, Gates went ahead on his own and formed the American Steel and Wire Company of Illinois. Quick success in the Cuban War created a buzz of optimism, investment flowed into the new company, and now Gates could buy more and more steel and wire companies, incorporating, in 1899, the U.S. Steel and Wire Company.[89] The new corporation had a capitalization of $90 million, and the world had never seen its like. The company dominated the entire steel and wire industry of the United States and hence the world. Gates was now as powerful as Morgan, a leader of American industry—a giant career built on the wounded bulls of San Antonio.[90]

Everywhere industries were centralizing. When trusts along the American lines could not be formed, companies reached together to form syndicates that oversaw production and prices. Consider Provoloka (Russian for "wire"), formed in Russia in 1908, controlling the entirety of Russian wire production from Poland to the Urals.[91] Also consider the much more important Deutsche Draht-Verband GmbH, formed in Düsseldorf "as the result of a contract entered into June 6, 1914 [whose] object was to improve the manufacture of wire and to further its sales at home and abroad."[92] (German foreign policy would soon take care of *that*.) But nothing compared to U.S. Steel and Wire. The Federal Trade Commission was justifiably enthusiastic: "In 1913 this company had 268 foreign agencies, in about 60 countries. It also had 40 foreign warehouses, situated in Antwerp, Johannesburg, Sydney, Copenhagen, Barcelona, Singapore, Valparaiso, Rio de Janeiro, and other places. It ordinarily had under charter 35 to 40 steamers for the transportation of its goods, which are sold as far north as Iceland and as far south as the Straits of Magellan and the South Sea Islands."[93]

Such concentrations were a typical feature of this age of capitalism, but so was barbed wire itself. The urge of the period was to concentrate, quite literally, to bring space under control. The initial urge then can be located on the prairies themselves, as entrepreneurs sought to bring land and cattle under control. This nineteenth-century capitalism was already based on the need for mass markets and mass products and therefore needed to have control on a vast scale. Hence the new kind of colonialism over an area, and the fencing of the plains. As the century ended, the fencers were fenced, and a few trusts reached everywhere—indeed, from Iceland to the South Seas. The entire structure of control over space—now global in nature—was firmly in position, and the ultimate locus of control was clearly seen to be at the center of capital, the Atlantic seaboard.

Maps 1 and 2 may be instructive in this respect. Map 1 indicates, in a rough way, the North American areas where barbed wire was chiefly distributed.[94] Map 2 indicates, with greater accuracy, the North American areas where barbed wire was chiefly produced.[95] We see, in this comparison, the phenomenon of spatial concentration. Because barbed wire is characterized by its relatively light weight, it makes economic sense to concentrate its production where the means of production are already available and then to distribute the product elsewhere. The main sites of production were the steel areas of Massachusetts, Pennsylvania, and Illinois; distribution was much wider and effectively covered the whole of the United States (and as the disproportionate role of seaports indicates, distribution was global). The light weight of barbed wire makes it into a cheap way of controlling space on the ground; it also makes it possible to produce this control over space from a few centers, so that Chicago and Pittsburgh, in this indirect way, come to control the space of America.

At this stage, two further comparisons should be made to complete this North American map. Upward, the financial and industrial control of Washburn and Moen, and then the American Steel and Wire Company, should be brought into the picture as well. Now we see the control of midwestern production by eastern capital, typical to this historical period.[96] Downward, and most important, one should add the animals themselves, all around the continent, ever more effectively surrounded and controlled by barbed wire. Now extend the picture globally, to appreciate the system just sketched, including arrangements such as those obtained with Felten and Guilleaume and with Johnson and Nephew. The resulting picture is that of the life of animals, throughout the globe, brought under human control through violence and pain, gain being extracted from this new form of control; and then control leads to control, until we reach the centers of control by capital, where violence and pain are no longer suffered or meted out, in places such as Mulheim, Germany; Manchester, England; and, above all, the American Northeast. It is unfortunate that Marx did not comment on this process, which

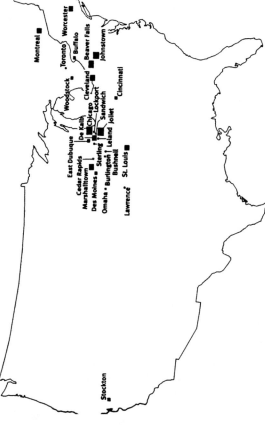

MAP 1 *Centers for delivery of barbed wire, United States, 1888.* Data from Roberts Wire Company, of Pittsburgh Pennsylvania, in the American Steel and Wire Company Archives (Baker MSS: 596 DcB 1119).

MAP 2 *Centers for production of barbed wire, United States, 1881.* Data from files relating to Washburn and Moen in the American Steel and Wire Company Archives (Baker MSS 596: DcC 827).

is perhaps a mere accident. Invented in 1874, barbed wire's economic role would become obvious only in Marx's last years. Barbed wire was destined to play a prominent role not so much in the theory of Marxism as in its practice (more on this in chapter 3).

Capitalist concentration itself, rather than the produce of the Great Plains, would be the true economic significance of barbed wire. The promise of the Great Plains gave a push to an industry of a certain tool of violence, and this industry gave a push to the concentration of capital. But the promise of the Great Plains remained deceptive. Of course, American capital did eventually develop intensive agriculture even in that arid land. Windmills brought water from beneath the surface; tractors tilled it over. In World War I and its aftermath, the production of Kansas, Oklahoma, and neighboring areas would be crucial in helping America to feed the world. But all the while, native vegetation was being destroyed, the soil overturned. Quite simply, the soil was not ready for intensive agriculture. Years of good rainfall helped the land from turning into dry dust, but when drought hit in the 1930s, the plains had already been denuded. Heavy winds always raged across the plains. Now they lifted up the soil, creating biblical clouds of black dust. Throughout the dry 1930s, these dust storms never completely ceased in the so-called Dust Bowl (an area encompassing parts of Kansas, Colorado, New Mexico, Texas, and Oklahoma). A storm could have a thick front of dust reaching up for a mile or more, weighing hundreds of tons per square mile, running across hundreds of miles on the open plains. Visibility disappeared, breathing was difficult—many people died of lung-related complaints—and everything, plants as well as animals, could be buried in the ensuing debris. The same would happen to people who had the bad luck to stray outdoors when a storm hit: "On March 15, 1935, a black blizzard struck Hays, Kansas, catching a seven-year-old boy away from home. The next morning a search party found him covered with dust and smothered. A hundred miles to the west, the same storm stranded a nine-year-old boy; a search party found him the next morning alive but tangled in barbed wire."[97] (We should recall Dorothy, another native of Kansas, having her own narrow escape.) In the traumatized ecology, rabbits suddenly proliferated, eating away the little produce that remained. Here, in the mid-southern part of the Great Plains, American colonization ensued in a terrible ecological blunder that in a sense never healed. Intensive efforts at soil conservation, as well as better luck with the rain, helped the Dust Bowl out in the 1940s, yet the area never did regain its place in the American economy.[98] Kansas, we can say, was a harbinger not of future development but of future underdevelopment. Throughout the Third World, through the twentieth century, modernism would bring the illusion of rapid development. The temptation would be to go down the path of an environmentally irresponsible monoculture, designed for the consumption of distant, rich lands. Early successes would typically lead to ecological and economic disasters. Unlike other Third World farmers, however, Kansans could vote for the U.S. Congress, and so they got their subsidies and somehow managed to extricate themselves from the legacy of the Dust Bowl.

So it is not in agriculture that the Great Plains formed a modern success story. Their significance lies in concentration, in control over space itself. This significance, however, is considerable, and it ushered in a special kind of modernity. For several decades, the plains were the prize of colonialism and an engine for historical change. At this cutting edge of history, barbed wire was created. By the end of the nineteenth century, the cutting edge of history was pulling away from the Great Plains, and barbed wire would soon make history elsewhere.

WHAT'S WRONG WITH ANIMAL RIGHTS

Of hounds, horses, and Jeffersonian happiness
By Vicki Hearne

Not all happy animals are alike. A Doberman going over a hurdle after a small wooden dumbbell is sleek, all arcs of harmonious power. A basset hound cheerfully performing the same exercise exhibits harmonies of a more lugubrious nature. There are chimpanzees who love precision the way musicians or fanatical housekeepers or accomplished hypochondriacs do; others for whom happiness is a matter of invention and variation—chimp vaudevillians. There is a rhinoceros whose happiness, as near as I can make out, is in needing to be trained every morning, all over again, or else he "forgets" his circus routine, and in this you find a clue to the slow, deep, quiet chuckle of his happiness and to the glory of the beast. Happiness for Secretariat is in his ebullient bound, that joyful length of stride. For the draft horse or the weight-pull dog, happiness is of a different shape, more awesome and less obviously intelligent. When the pulling horse is at its most intense, the animal goes into himself, allocating all of the educated power that organizes his desire to dwell in fierce and delicate intimacy with that power, leans into the harness, and MAKES THAT SUCKER MOVE.

If we are speaking of human beings and use the phrase "animal happiness," we tend to mean something like "creature comforts." The emblems of this are the golden retriever rolling in the grass, the horse with his nose deep in the oats, the kitty by the fire. Creature comforts are important to animals—"Grub first, then ethics" is a motto that would describe many a wise Labrador retriever, and I have a pit bull named Annie whose continual quest for the perfect pillow inspires her to awesome feats. But there is something more to animals, a capacity for satisfactions that come from work in the fullest sense—what is known in philosophy and in this country's Declaration of Independence as "happiness." This is a sense of

Vicki Hearne is a contributing editor of Harper's Magazine, *an animal trainer, and the author of* Bandit: Dossier of a Dangerous Dog, *which will be published in November by HarperCollins. Her last piece for* Harper's Magazine, *"Beware of the Dog!" appeared in the February 1986 issue.*

WORK IS THE FOUNDATION OF THE HAPPINESS A TRAINER AND AN ANIMAL DISCOVER TOGETHER

personal achievement, like the satisfaction felt by a good wood-carver or a dancer or a poet or an accomplished dressage horse. It is a happiness that, like the artist's, must come from something within the animal, something trainers call "talent." Hence, it cannot be imposed on the animal. But it is also something that does not come *ex nihilo*. If it had not been a fairly ordinary thing, in one part of the world, to teach young children to play the pianoforte, it is doubtful that Mozart's music would exist.

Happiness is often misunderstood as a synonym for pleasure or as an antonym for suffering. But Aristotle associated happiness with ethics—codes of behavior that urge us toward the sensation of getting it right, a kind of work that yields the "click" of satisfaction upon solving a problem or surmounting an obstacle. In his *Ethics*, Aristotle wrote, "If happiness is activity in accordance with excellence, it is reasonable that it should be in accordance with the highest excellence." Thomas Jefferson identified the capacity for happiness as one of the three fundamental rights on which all others are based: "life, liberty, and the pursuit of happiness."

I bring up this idea of happiness as a form of work because I am an animal trainer, and work is the foundation of the happiness a trainer and an animal discover together. I bring up these words also because they cannot be found in the lexicon of the animal-rights movement. This absence accounts for the uneasiness toward the movement of most people, who sense that rights advocates have a point but take it too far when they liberate snails or charge that goldfish at the county fair are suffering. But the problem with the animal-rights advocates is not that they take it too far; it's that they've got it all wrong.

Animal rights are built upon a misconceived premise that rights were created to prevent us from unnecessary suffering. You can't find an animal-rights book, video, pamphlet, or rock concert in which someone doesn't mention the Great Sentence, written by Jeremy Bentham in 1789. Arguing in favor of such rights, Bentham wrote: "The question is not, Can they *reason*? nor, can they *talk*? but, can they suffer?"

The logic of the animal-rights movement places suffering at the iconographic center of a skewed value system. The thinking of its proponents—given eerie expression in a virtually sado-pornographic sculpture of a tortured monkey that won a prize for its compassionate vision—has collapsed into a perverse conundrum. Today the loudest voices calling for—demanding—the destruction of animals are the humane organizations. This is an inevitable consequence of the apotheosis of the drive to relieve suffering: Death is the ultimate release. To compensate for their contradictions, the humane movement has demonized, in this century and the last, those who made animal happiness their business: veterinarians, trainers, and the like. We think of Louis Pasteur as the man whose work saved you and me and your dog and cat from rabies, but antivivisectionists of the time claimed that rabies increased in areas where there were Pasteur Institutes.

An anti-rabies public-relations campaign mounted in England in the 1880s by the Royal Society for the Prevention of Cruelty to Animals and other organizations led to orders being issued to club any dog found not wearing a muzzle. England still has her cruel and unnecessary law that requires an animal to spend six months in quarantine before being allowed loose in the country. Most of the recent propaganda about pit bulls—the crazy claim that they "take hold with their front teeth while they chew away with their rear teeth" (which would imply, incorrectly, that they have double jaws)—can be traced to literature published by the Humane Society of the United States during the fall of 1987 and earlier. If your neighbors want your dog or horse impounded and destroyed because he is a nuisance—say the dog barks, or the horse attracts flies—it will be the local Humane Society to whom your neighbors turn for action.

In a way, everyone has the opportunity to know that the history of the humane movement is largely a history of miseries, arrests, prosecutions, and death. The Humane Society is the pound, the place with the decompression chamber or the lethal injections. You occasionally find worried letters about this in Ann Landers's column.

Animal-rights publications are illustrated largely with photographs of two kinds of animals—"Helpless Fluff" and "Agonized Fluff," the two conditions in which some people seem to prefer their animals, because any other version of an animal is too complicated for propaganda. In the introduction to his book *Animal Liberation*, Peter Singer says somewhat smugly that he and his wife have no animals and, in fact, don't much care for them. This is offered as evidence of his objectivity and ethical probity. But it strikes me as an odd, perhaps obscene, underpinning for an ethical project that encourages university and high school students to cherish their ignorance of, say, great bird dogs as proof of their devotion to animals.

THE WILD IS NOT
A SUFFERING-FREE ZONE
OR ALL THAT FROLICSOME
A LOCATION

I would like to leave these philosophers behind, for they are inept connoisseurs of suffering who might revere my Airedale for his capacity to scream when subjected to a blowtorch but not for his wit and courage, not for his natural good manners that are a gentle rebuke to ours. I want to celebrate the moment not long ago when, at his first dog show, my Airedale, Drummer, learned that there can be a public place where his work is respected. I want to celebrate his meticulousness, his happiness upon realizing at the dog show that no one would swoop down upon him and swamp him with the goo-goo excesses known as the "teddy-bear complex" but that people actually got out of his way, gave him room to work. I want to say, "There can be a six-and-a-half-month-old puppy who can care about accuracy, who can be fastidious, and whose fastidiousness will be a foundation for courage later." I want to say, "Leave my puppy alone!"

I want to leave the philosophers behind, but I cannot, in part because the philosophical problems that plague academicians of the animal-rights movement are illuminating. They wonder, do animals have rights or do they have interests? Or, if these rightists lead particularly unexamined lives, they dismiss that question as obvious (yes, of course, animals have rights, prima facie) and proceed to enumerate them, James Madison style. This leads to the issuance of bills of rights—the right to an environment, the right not to be used in medical experiments—and other forms of trivialization.

The calculus of suffering can be turned against the philosophers of festering flesh, even in the case of food animals, or exotic animals who perform in movies and circuses. It is true that it hurts to be slaughtered by man, but it doesn't hurt nearly as much as some of the cunningly cruel arrangements meted out by "Mother Nature." In Africa, 75 percent of the lions cubbed do not survive to the age of two. For those who make it to two, the average age at death is ten years. Asali, the movie and TV lioness, was still working at age twenty-one. There are fates worse than death, but twenty-one years of a close working relationship with Hubert Wells, Asali's trainer, is not one of them. Dorset sheep and polled Herefords would not exist at all were they not in a symbiotic relationship with human beings.

A human being living in the "wild"—somewhere, say, without the benefits of medicine and advanced social organization—would probably have a life expectancy of from thirty to thirty-five years. A human being living in "captivity"—in, say, a middle-class neighborhood of what the Centers for Disease Control call a Metropolitan Statistical Area—has a life expectancy of seventy or more years. For orangutans in the wild in Borneo and Malaysia, the life expectancy is thirty-five years; in captivity, fifty years. The wild is not a suffering-free zone or all that frolicsome a location.

The questions asked by animal-rights activists are flawed, because they are built on the concept that the origin of rights is in the avoidance of

WHAT KIND OF THING CAN MY AIREDALE, DRUMMER, HAVE KNOWLEDGE OF?

suffering rather than in the pursuit of happiness. The question that needs to be asked—and that will put us in closer proximity to the truth—is not, do they have rights? or, what are those rights? but rather, what is a right?

Rights originate in committed relationships and can be found, both intact and violated, wherever one finds such relationships—in social compacts, within families, between animals, and between people and nonhuman animals. This is as true when the nonhuman animals in question are lions or parakeets as when they are dogs. It is my Airedale whose excellencies have my attention at the moment, so it is with reference to him that I will consider the question, what is a right?

When I imagine situations in which it naturally arises that A defends or honors or respects B's rights, I imagine situations in which the relationship between A and B can be indicated with a possessive pronoun. I might say, "Leave her alone, she's my daughter" or, "That's what she wants, and she is my daughter. I think I am bound to honor her wants." Similarly, "Leave her alone, she's my mother." I am more tender of the happiness of my mother, my father, my child, than I am of other people's family members; more tender of my friends' happinesses than your friends' happinesses, unless you and I have a mutual friend.

Possession of a being by another has come into more and more disrepute, so that the common understanding of one person possessing another is slavery. But the important detail about the kind of possessive pronoun that I have in mind is reciprocity: If I have a friend, she has a friend. If I have a daughter, she has a mother. The possessive does not bind one of us while freeing the other; it cannot do that. Moreover, should the mother reject the daughter, the word that applies is "disown." The form of disowning that most often appears in the news is domestic violence. Parents abuse children; husbands batter wives.

Some cases of reciprocal possessives have built-in limitations, such as "my patient / my doctor" or "my student / my teacher" or "my agent / my client." Other possessive relations are extremely limited but still remarkably binding: "my neighbor" and "my country" and "my president."

The responsibilities and the ties signaled by reciprocal possession typically are hard to dissolve. It can be as difficult to give up an enemy as to give up a friend, and often the one becomes the other, as though the logic of the possessive pronoun outlasts the forms it chanced to take at a given moment, as though we were stuck with one another. In these bindings, nearly inextricable, are found the origin of our rights. They imply a possessiveness but also recognize an acknowledgment by each side of the other's existence.

The idea of democracy is dependent on the citizens' having knowledge of the government; that is, realizing that the government exists and knowing how to claim rights against it. I know this much because I get mail from the government and see its "representatives" running about in uniforms. Whether I actually have any rights in relationship to the government is less clear, but the idea that I do is symbolized by the right to vote. I obey the government, and, in theory, it obeys me, by counting my ballot, reading the *Miranda* warning to me, agreeing to be bound by the Constitution. My friend obeys me as I obey her; the government "obeys" me to some extent, and, to a different extent, I obey it.

What kind of thing can my Airedale, Drummer, have knowledge of? He can know that I exist and through that knowledge can claim his happinesses, with varying degrees of success, both with me and against me. Drummer can also know about larger human or dog communities than the one that consists only of him and me. There is my household—the other dogs, the cats, my husband. I have had enough dogs on campuses to know that he can learn that Yale exists as a neighborhood or village. My older dog, Annie, not only knows that Yale exists but can tell Yalies from townies, as I learned while teaching there during labor troubles.

Dogs can have elaborate conceptions of human social structures, and even of something like their rights and responsibilities within them, but these conceptions are never elaborate enough to construct a rights relationship between a dog and the state, or a dog and the Humane Society. Both of these are concepts that depend on writing and memoranda, officers in uniform, plaques and seals of authority. All of these are literary constructs, and all of them are beyond a dog's ken, which is why the mail carrier who doesn't also happen to be a dog's friend is forever an intruder—this is why dogs bark at mailmen.

It is clear enough that natural rights relations can arise between people and animals. Drummer, for example, can insist, "Hey, let's go outside and do something!" if I have been at my computer several days on end. He can both refuse to accept various of my suggestions and tell me when he fears for his life—such as the time when the huge, white flapping flag appeared out of nowhere, as it seemed to him, on the town green one evening when we were working. I can (and do) say to him either, "Oh, you don't have to worry about that" or, "Uh oh, you're right, Drum, that guy looks dangerous." Just as the government and I—two different species of organism—have developed improvised ways of communicating, such as the vote, so Drummer and I have worked out a number of ways to make our expressions known. Largely through obedience, I have taught him a fair amount about how to get responses from me. Obedience is reciprocal; you cannot get responses from a dog to whom you do not respond accurately. I have enfranchised him in a relationship to me by educating him, creating the conditions by which he can achieve a certain happiness specific to a dog, maybe even specific to an Airedale, inasmuch as this same relationship has allowed me to plumb the happiness of being a trainer and writing this article.

Instructions in this happiness are given terms that are alien to a culture in which liver treats, fluffy windup toys, and miniature sweaters are confused with respect and work. Jack Knox, a sheepdog trainer originally from Scotland, will shake his crook at a novice handler who makes a promiscuous move to praise a dog, and will call out in his Scottish accent, "Eh! Eh! Get back, get BACK! Ye'll no be abusin' the dogs like that in my clinic." America is a nation of abused animals, Knox says, because we are always swooping at them with praise, "no gi'ing them their freedom." I am reminded of Rainer Maria Rilke's account in which the Prodigal Son leaves—has to leave—because everyone loves him, even the dogs love him, and he has no path to the delicate and fierce truth of himself. Unconditional praise and love, in Rilke's story, disenfranchise us, distract us from what truly excites our interest.

In the minds of some trainers and handlers, praise is dishonesty. Paradoxically, it is a kind of contempt for animals that masquerades as a reverence for helplessness and suffering. The idea of freedom means that you do not, at least not while Jack Knox is nearby, helpfully guide your dog through the motions of, say, herding over and over—what one trainer calls "explainy-wainy." This is rote learning. It works tolerably well on some handlers, because people have vast unconscious minds and can store complex pre-programmed behaviors. Dogs, on the other hand, have almost no unconscious minds, so they can learn only by thinking. Many children are like this until educated out of it.

If I tell my Airedale to sit and stay on the town green, and someone comes up and burbles, "What a pretty thing you are," he may break his stay to go for a caress. I pull him back and correct him for breaking. Now he holds his stay because I have blocked his way to movement but not because I have punished him. (A correction blocks one path as it opens another for desire to work; punishment blocks desire and opens nothing.) He holds his stay now, and—because the stay opens this possibility of work, new to a heedless young dog—he watches. If the person goes on

OUR CULTURE CONFUSES LIVER TREATS AND MINIATURE SWEATERS WITH RESPECT FOR ANIMALS

I AM THE ONLY ONE WHO CAN OWN UP TO MY AIREDALE'S INALIENABLE RIGHTS

talking, and isn't going to gush with praise, I may heel Drummer out of his stay and give him an "Okay" to make friends. Sometimes something about the person makes Drummer feel that reserve is in order. He responds to an insincere approach by sitting still, going down into himself, and thinking, "This person has no business pawing me. I'll sit very still, and he will go away." If the person doesn't take the hint from Drummer, I'll give the pup a little backup by saying, "Please don't pet him, he's working," even though he was not under any command.

The pup reads this, and there is a flicker of a working trust now stirring in the dog. Is the pup grateful? When the stranger leaves, does he lick my hand, full of submissive blandishments? This one doesn't. This one says nothing at all, and I say nothing much to him. This is a working trust we are developing, not a mutual-congratulation society. My backup is praise enough for him; the use he makes of my support is praise enough for me.

Listening to a dog is often praise enough. Suppose it is just after dark and we are outside. Suddenly there is a shout from the house. The pup and I both look toward the shout and then toward each other: "What do you think?" I don't so much as cock my head, because Drummer is growing up, and I want to know what he thinks. He takes a few steps toward the house, and I follow. He listens again and comprehends that it's just Holly, who at fourteen is much given to alarming cries and shouts. He shrugs at me and goes about his business. I say nothing. To praise him for this performance would make about as much sense as praising a human being for the same thing. Thus:

A. What's that?
B. I don't know. [Listens] Oh, it's just Holly.
A. What a goooooood human being!
B. Huh?

This is one small moment in a series of like moments that will culminate in an Airedale who on a Friday will have the discrimination and confidence required to take down a man who is attacking me with a knife and on Saturday clown and play with the children at the annual Orange Empire Dog Club Christmas party.

People who claim to speak for animal rights are increasingly devoted to the idea that the very keeping of a dog or a horse or a gerbil or a lion is in and of itself an offense. The more loudly they speak, the less likely they are to be in a rights relation to any given animal, because they are spending so much time in airplanes or transmitting fax announcements of the latest Sylvester Stallone anti-fur rally. In a 1988 *Harper's* forum, for example, Ingrid Newkirk, the national director of People for the Ethical Treatment of Animals, urged that domestic pets be spayed and neutered and ultimately phased out. She prefers, it appears, wolves—and wolves someplace else—to Airedales and, by a logic whose interior structure is both emotionally and intellectually forever closed to Drummer, claims thereby to be speaking for "animal rights."

She is wrong. I am the only one who can own up to my Airedale's inalienable rights. Whether or not I do it perfectly at any given moment is no more refutation of this point than whether I am perfectly my husband's mate at any given moment refutes the fact of marriage. Only people who know Drummer, and whom he can know, are capable of this relationship. PETA and the Humane Society and the ASPCA and the Congress and NOW—as institutions—do have the power to affect my ability to grant rights to Drummer but are otherwise incapable of creating conditions or laws or rights that would increase his happiness. Only Drummer's owner has the power to obey him—to obey who he is and what he is capable of—deeply enough to grant him his rights and open up the possibility of happiness. ∎

CHAPTER TWO

Prize Pets

In 1911 Judith Neville Lytton's *Toy Dogs and Their Ancestors* took on what was known as the doggy world of England.[1] Lytton was a knowledgeable and opinionated young dog breeder, tennis champion, and general animal fancier, who later wrote extensively about thoroughbred horses. Her aristocratic background—she was a baroness in her own right and the wife of an earl—distinguished her from most serious fanciers of pedigreed dogs. Similarly, *Toy Dogs* did not belong to the ordinary run of breed books, which cozily rehearsed the accepted standards for show dogs, spotlighted a few successful kennels, then dispensed received wisdom about feeding, treatment of minor illnesses, and similar quotidian concerns. Lytton relegated these routine items to the end of her richly illustrated volume. She devoted the initial chapters to a systematic assault on the contemporary toy dog fancy, mounting her attack in terms that could have been applied to any group of pedigreed dogs and their admirers.

In Lytton's view giving detailed practical advice to owners and breeders was putting the cart before the horse. Before fanciers devoted effort and expense to raising and showing dogs, they had to ensure that the animals merited the outlay. Although an elaborate structure for assessing and certifying dogs had developed over the previous half-century—a pyramid based on local and breed-specific clubs and shows, culminating in the national, multibreed Kennel Club—she felt that this struc-

ture was resting on sand, or worse. It was, she charged, designed to enforce standards that had no basis in nature or aesthetics but reflected the ignorant, self-interested caprices of fanciers who wished to boost the prestige of their own stock. Not even the accepted breed categories survived her scrutiny. She used strong language to emphasize the seriousness of the situation. To outsiders, she charged, "the present judging system appears ridiculous and contemptible."[2]

She based her criticism, in part, on obvious defects of animals with classy pedigrees: they were physically unsound, they had sparse coats and ugly expressions, and they were excessively timid, sluggish, and idiotic. These faults had only to be pointed out to be acknowledged. Any breeder with even a smattering of Darwin would have agreed that art should follow nature in preferring the strong and beautiful to the weak and grotesque. But this was only the beginning of Lytton's complaint. Much of what displeased her about Edwardian toy spaniels—square jaws, black and tan coats, and relatively stocky physiques—seemed rather neutral on that stern scale. Such characteristics revealed what Lytton considered the most unforgivable flaw of the toy spaniels of her day: that they had diverged from their historical prototypes. Although their ancestors could be certified for many generations, they did not, in her opinion, resemble the dogs that had established their ostensible breed. As a result, she insisted, their pedigrees were fundamentally fraudulent, however long and well documented.[3]

There was no question that the toy spaniel had a venerable history. The names of its two major sub-breeds took it back at least to the Restoration: the King Charles (named for Charles II) and the Blenheim (after the palace built by the first duke of Marlborough). The breed, or one closely related, may also have flourished under the name of "comforter" during the reign of Elizabeth I.[4] In Lytton's view the toy spaniels she saw around her had diverged from the original Stuart or Tudor pattern in color, size, and facial configuration. The scarcity of concrete information about the pre-nineteenth-century history of the breed made these assertions difficult to prove, but she marshaled an eclectic variety of evidence, including pictures as well as written documents, to demonstrate that "the present square-jawed, heavy, noseless type was introduced ... no earlier than ... 1840"; that it was, indeed, a "modern fake"; that "the whole red variety ... cannot be traced back more than eighty years"; and that "the present standard and scale of points had

PRESTIGE AND PEDIGREE

apparently no foundation earlier than 1885 or 1887." In sum, she asserted, "there is a hopeless confusion in the naming of breeds and in the type desired."[5]

Lytton's criticism struck at the heart of the dog fancy, focused as it was on a series of finely graded differentiations, which functioned both to establish the unique character of each breed and to assess the relative excellence of dogs of the same breed. Firmly grounded in the animals' physical attributes, which were endlessly rehearsed in manuals and concretely exemplified in show after show, this elaborate system of categories metaphorically expressed the hopes and fears of fanciers about issues like social status and the need for distinctions between classes. Unlike Lytton, most late Victorian dog fanciers belonged to the urban business and professional classes. To many of them, the figurative dimension of dog fancying may have been the more important; it offered a vision of a stable, hierarchical society, where rank was secure and individual merit, rather than just inherited position, appreciated The content of fancying categories thus became a by-product of the process of assigning ranks and orders, and one powerful criterion for a specification was whether it made this process easier. By analyzing the physical consequences of this policy, Lytton also undermined the rhetorical manipulations that masked it Toy spaniels had lost their historical authenticity as well as their beauty and vigor at the hands of fanciers who had disingenuously proclaimed that they were protecting and improving the breed.

Class Barriers

Lytton assaulted the dog fancying establishment as both a radical and a conservative. Her method was that of a Young Turk, but her accusations implicitly labeled adherents of the entrenched bureaucracy as arrivistes. They cherished the letter of tradition, she claimed, with their punctilious recording of pedigrees, while disregarding its spirit. Although they might place themselves within the ancient British tradition of dog breeding, she insinuated, only aristocratic fanciers like herself, who strove to maintain old values and old stock, really belonged there. Although Lytton did not go so far as to say so, this very association of dog breeding with the social elite was one of the things that attracted the fanciers whose self-interested manipulations she deplored Adopt-

Prize Pets

ing an upper-class avocation was a way of reinforcing their own social position. And the dog fancy was perhaps uniquely open to people of moderate means—certainly much more so than high stock breeding. Although dog breeding was not apt to be a profitable hobby, it did not require either great wealth or broad acres. A beginner could invest a fortune—one Mr. Stephens of Action spent £2,000 in setting up a fox terrier kennel—but it was generally agreed that £25, wisely laid out, should suffice.[6]

The British had owned dogs from the beginning of recorded history, but the relation of most Victorian fanciers to their animals, kept purely for companionship and amusement, was rather new, especially outside the highest social ranks. The invading Romans had described animals identified by nineteenth-century experts as mastiffs, and King Alfred's laws included fines to be levied on the owners of dogs that killed or maimed people. The subjects of these ancient references were doubtless working animals, however, as were the medieval ancestors of later hounds and gun dogs.[7] Unmistakable pets first appeared in the middle ages, as the playthings of courtiers and members of privileged religious orders. Episcopal authorities tried repeatedly to suppress petkeeping among monks and nuns, with little success, but the practice did not spread among ordinary citizens.[8] Although working dogs were ubiquitous in the Renaissance—they turned cooking spits, pulled carts, herded sheep, retrieved game, baited wild animals, and competed in sporting events—pet dogs remained the province of the upper classes, particularly of their female members. King Charles II may have been the first man to declare a public passion for them. When his pets were stolen, as seems to have happened with some frequency, he was inconsolable; once he advertised in a newspaper for a favorite's return. His brother, James II, shared this fondness, as, apparently, did his successors, William and Mary, during whose reign the pug, like William a native of the Netherlands, became established among the English aristocracy At about the same time, dogs with elite connections, both pets and sporting animals, began to sit for formal portraits[9]

Despite these exalted pacesetters, it was almost a century before petkeeping became respectable among ordinary citizens. To some extent, the change paralleled the increased public indulgence of the softer emotions during the last part of the eighteenth century and found literary expression in, for example, Christopher Smart's apostrophe to his

PRESTIGE AND PEDIGREE

cat Jeoffry.[10] Although the most extravagant avowals of attachment to pets were still reserved for the upper classes, by the early nineteenth century such effusions were more likely to be greeted with general sympathy than with disapprobation or derision. When, in 1808, Lord Byron (Lytton's great-grandfather) buried his Newfoundland dog Boatswain "within . . . the precincts of the sacred Abbey of Newstead," many of his less distinguished countrymen shared his sense that a dog offered greater loyalty and affection than any human friend or servant—or, as Byron put it on the monument, "all the Virtues of Man without his vices."[11] Increasingly, they felt the same way about their own animals. By the middle of the nineteenth century what has been called the Victorian cult of pets was firmly established. *Punch* frequently satirized the foolishness of dog lovers who fed their pets from the table, dressed them in elaborate outfits, and allowed them to inconvenience human members of the household.

Love was not, however, the whole story. The intensifying attachment of members of the middle classes to their animals could also be measured in cash. Entrepreneurs quickly exploited a range of new commercial opportunities. By mid-century there were approximately twenty thousand London street traders who dealt in live animals, and at least a dozen who specialized in the brass collars, priced from 6d. to 3s. apiece, sported by most respectable Victorian dogs. On the shady side, professional dog stealers would abscond with a cherished animal, then offer to restore it for a price. In 1844 individual Londoners paid from £2 to £50 to ransom favorite pets.[12] More formally marketed were such kennel care products as Spratts Patent Meat "Fibrine" Dog Cakes ("As supplied to the Royal Kennels"), Ashworth's Patent "Metallic Comb-Brush," and, at £10 10s., Boulton and Paul's Dog's House and Yard Combined. At the top of the line a few merchants indulged clients who wished to pay 30s. to have their monogram clipped into their dog's fur, or £5 for a satin wedding coat, or up to £60 for collars and other ornaments of gold and silver.[13]

The publishing industry also catered to the increasing audience of middle-class dog fanciers. Before the nineteenth century, books about dogs were few and far between, and most simply rehashed the curious information offered in Johannus Caius's *De Canibus Britannicus*, published in Latin in 1570 and in translation, as *Of Englishe Dogges*, in 1576. Caius's treatise was an annotated list of Tudor dog types, compiled as a favor

Prize Pets

to the Swiss bestiarist Konrad Gesner.[14] An expanded market inspired a sudden stream of dog books beginning with Sydenham Edwards's handsomely illustrated *Cynographia Britannica*, which was issued in parts between 1800 and 1805. These volumes reflected both the emotional and the material concerns of genteel pet owners. Whereas earlier dog literature seemed simply a specialized branch of natural history, the new books included not only descriptions of the dogs' physical and moral characteristics, but a selection of heartwarming and enlightening anecdotes. In addition, like the animals they described, most were luxury items, suitable for conspicuous display.[15]

The cherished animals themselves carried impressive price tags, which increased as the dog fancy flourished. Indeed, Charles Rotherham, the veterinarian who attended Queen Victoria's kennel, attributed an absolute rise in the canine population of London between approximately 1865 and 1887 to the snowballing value of purebred dogs. And in turn, according to one journal of the fancy, the "high prices paid nowadays" showed "the progress of the canine race." By 1891 champion collies and St. Bernards sold for £1,000, and hopeful buyers offered £375 for a prize fox terrier and £250 for a particularly distinguished King Charles spaniel.[16] Price was a sensitive indicator of differences among animals. Pedigreed but undistinguished specimens of popular breeds could be had for much less—about 3 guineas if they were merely of pet quality, and a minimum of £10 if "sufficiently perfect" to win in the least competitive shows.[17]

Such carefully chronicled expenditure referred ultimately to the status of the owner rather than that of the dog. Despite its genuinely sentimental roots, much middle-class petkeeping was shadowed by similar motivations. After all, the maintenance of idle animals was a custom borrowed from the upper echelons of society and constituted a metonymic attempt at assimilation, the elaborate certification and registration of pedigreed animals was hardly designed to guarantee their emotional qualities. The incorporation of dogs into the rhetoric of social aspiration did not go unnoticed by those whose practices were being appropriated. Lytton was not the first aristocratic fancier to draw a line between traditional fancying and that introduced in the course of the nineteenth century, although she was rather atypical in choosing to focus her concern on a pet breed. Members of the elite more frequently tried to draw a line between their companion animals—the sporting dogs traditionally associated with rural life—and the pet dogs, designed simply for human pleasure, that urban fanciers favored. As early as 1824, the anonymous but red-blooded author of *The Complete Dog Fancier, or, General History of Dogs* had announced that, his title notwithstanding, "those [dogs] which some fond ladies make their daily passtime, have no business in these pages."[18] His animosity was echoed more mildly in the content of many contemporary volumes of canine appreciation. Until about 1840, most of them were ostensibly addressed to sportsmen and paid primary attention to hounds and gun dogs; but afterward the burgeoning literature gave at least equal attention to the pet breeds typically kept by prosperous urban dog owners.

Rural fanciers were no better pleased when their favorite breeds began to appear as urban pets and in the show ring. It was common knowledge that "Masters of Fox Hounds abominate dog shows"; one sportsman suggested that they "have occasioned more mischief than years on years will serve to eradicate." They feared that animals bred as show dogs or pets would be useless in the field. To address these fears the Kennel Club instituted competitive field trials for sporting breeds like setters and pointers, the results of which supplemented show bench rankings by measures of actual performance. Even "Stonehenge," a crusty critic and perhaps the most highly respected sporting journalist of his time, admitted that this countermeasure had been effective. Yet the sporting interest continued to air its contempt for pet dog breeders.

After almost half a century of formal dog shows, the author of a manual for dog owners noted that "the sportsman will as a rule have nothing to do with the fancier's production."[19] Correctly interpreting the aspirations figured in these indulged animals, gentry fanciers were defending their turf as well as the utility of their animals.

Middle-class fanciers did not reciprocate this hostility. On the contrary, they were pleased to be associated with their betters, even if at cross-purposes. The higher the rank of a breeder, the more enthusiastic the public reception was likely to be. Titles were thinly scattered among the ranks of fanciers, and any aristocrat who became serious about showing was guaranteed effusive appreciation. Readers of the *British Fancier Annual Review*, for example, were treated to profiles of elegant kennels like those of the Prince of Wales at Sandringham and the duchess of Newcastle at Clumber House; the *Kennel Review* featured the Queen's favorite pets (collies, terriers, and a dachshund.) In 1896 the

Ladies' Kennel Journal published a photograph album entitled *Notable Dogs of the Year and Their Owners*. Some photographs included just the owner, some just the dog, and some both, but all were arranged according to the owner's rank. The opening section included the child Queen of the Netherlands and her charming but undistinguished Irish setters, as well as a large selection of Queen Victoria's dogs and many members of the royal family. A string of titled and honorable women followed and then, finally, the merely genteel fanciers and their pets. The less exalted the owner's social standing, the more likely was her pet to be a champion.[20]

When very high rank was at issue, some fanciers were willing to sacrifice even the meritocratic competition that structured their activities. The participation of royalty was especially coveted. Charles Lane (self-styled "breeder, exhibitor, judge") reported several occasions on which he was approached by members of show committees, who wished to ensure that the Queen's entries ended up in the prize list. But the stalwart judge stood for principle rather than excessive deference. He protested that "although I will not admit Her Majesty has a more loyal or devoted subject than myself, I am here in a public capacity as a judge; the royal dogs had to be judged "on their merits," like all the others. The resulting impartial awards, according to Lane, "caused general satisfaction," and he remained certain that they would "have been approved by Her Majesty ... if the circumstances came to be known at the palace."[21]

Elite patronage could boost the stock of a breed, as fanciers strove to identify their own tastes with those of their social superiors. Collies, Queen Victoria's favorite breed, were the most conspicuous beneficiaries of such preference; the sovereign's partiality also helped popularize Pomeranians. The pug, out of favor for much of the nineteenth century, was revived "thanks to the care of Lord Willoughby" and other aristocratic admirers. A foreign variant, the black pug, caught on quickly after its initial public appearance in 1886; its first English owners included Lord Londonderry, Lady Brassy, and the Queen.[22] The exalted rank of admirers might rub off on a breed, which could then pass it on to more ordinary fanciers. Pugs were often characterized as "aristocratic," as were collies, bloodhounds, borzois, and deerhounds, among others.[23] Once established as aristocrats, dogs received treatment commensurate with their rank. Fancy periodicals regularly featured profiles and portraits of distinguished champions. The *Fox-Terrier*

Chronicle, the only nineteenth-century periodical devoted to a single breed of dogs, covered the terrier elite the way that newspapers and other periodicals covered human high society. Issues might occasionally carry "Portraits of Fox Terrier Men," but the staple departments included "Portraits of Celebrated Terriers," "Biographies of Celebrated Dogs," "Gossip," "Visits" (a euphemism for matings), and "Debutantes" (dogs making their first dog show appearances).

Like people, however, dogs occupied the full range of social ranks; an English sportsman on safari described his canine retinue as including "dogs of high and low degree, from the purebred English greyhound to the Kaffir cur."[24] Most fanciers were as eager to dissociate themselves from the common people as to ally themselves with the aristocracy. Vulgar associations could preclude a breed's acceptance in genteel circles. Although the whippet, which had long been popular among laborers in Yorkshire and Lancashire, was recognized by the *Stud Book* in 1892, Rawdon Lee doubted that it would become popular. The pedigrees of even the listed specimens were not well documented, and "its surroundings have not, as a rule, been of the highest in the social scale."[25] In certain cases, nevertheless, genteel fanciers were willing to adopt animals from humble backgrounds, usually when they could view their previous owners as temporary custodians. Thus the fact that toy spaniels were bred in the East End of London as well as at Blenheim Palace did not impair their popularity. And, apparently, the readers of the *Sportsman's Journal and Fancier's Guide* did not scruple to accept the advice offered in January 1879: "The distress among the colliers in South Yorkshire ... combined with their inability to part with their pets. Dogs ... of no small value, can be had for nothing."[26]

Ordinarily, class barriers were rigidly observed, even, in the view of some experts, by the animals themselves. Edward Jesse, author of the compendious *Anecdotes of Dogs*, claimed that what he called the Irish wolfdog could identify descendants of the ancient kings of Ireland, on which *Punch* commented, "reader, if you can swallow that you can of course bolt the whole Book of Jesse without wincing."[27] Although he was a zoologist specializing in animal behavior, George Romanes could be similarly credulous. He retailed the story of a retriever who, while wandering alone, struck up an acquaintance with a rat-catcher and his cur, but as soon as his master appeared abruptly "cut" his new friends.

According to Romanes, this proved that dogs understood "the idea of caste."[28] In general, however, fanciers did not rely on the perception of their animals to ensure that they would keep suitable company. Although there were knowledgeable dog breeders at every level of English society, the fancying establishment was carefully segregated by class. Many dog clubs sponsored auxiliaries for less genteel fanciers with lower membership fees and reduced privileges. Similarly, some shows levied cut-rate registration charges on dogs entered in special workingmen's classes, so that even dogs of the same breed competed only with their social equals.

If association with the lower orders compromised the standing of even pedigreed animals, dogs without breed standing were unquestionably beyond the pale. As the stock of well-descended animals rose during the nineteenth century, that of commoners fell. Most of the dogs that Caius had described in the sixteenth century would probably have been lumped together by Victorian fanciers in the catch-all class of mongrels or curs. It was a class about which they had little good to say. According to one expert, its members did "ninety per cent of all the mischief that the canine community is charged with committing"; another stated that "old dogs of a poor breed often become very dirty. They were "useless" and "miserable"—nothing but "rubbish." Any lapse from purity was enough to consign an animal to this category. Prospective dog owners were admonished that "your half-mongrel Pomeranian is a perfect little brute—a heel-biter and a coward" and that "there cannot be an uglier, more selfish little beast of a dog ... than a mongrel or badly-bred pug." At best such half-castes (although probably not half-pugs) might be useful and loyal to "the farmer and the grazier."[29]

But it behooved more elevated members of society to avoid "vulgar companionship." Manual after manual warned that a careless choice of pet could signal the owner's lack of distinction and discrimination. Everett Millais, a highly respected writer on canine subjects, generously speculated that such gaffes might merely reflect "want of knowledge"; the explanation could not be a mean financial one, he suggested, because "it is an old maxim that it costs no more to keep a purebred dog than to keep a mongrel." But most observers felt that ownership of mongrels revealed latent commonness, especially since, as the *Dog Owners' Annual* crowed in 1890, the general tone of pet dogs had greatly improved. "Those of the past," it reported, "were, by comparison, little better than mongrels, whilst at the present time we constantly see dogs of all kinds of the purest type." With such competition, the same periodical warned several years later, "nobody now who is anybody can afford to be followed about by a mongrel dog."[30]

The Institutionalization of the Dog Fancy

If keeping a well-bred dog metonymically allied its owner with the upper ranges of society, then the elaborate structure of pedigree registration and show judging metaphorically equated owner with elite pet. The institutions that defined the dog fancy projected an obsessively detailed vision of a stratified order which sorted animals and, by implication, people into snug and appropriate niches. Dog breeds were split and split again to produce categories in which competitive excellence could be determined. Aficionados of each breed developed a set of points prescribing the ideal toward which breeders should direct their efforts, and these ideals were publicly ratified and enforced in dog shows, which offered a dizzying range of classes and then abstracted from them a carefully calibrated hierarchy of animals, ranging from those who did not place even in their sub-breed category to the best of show.

Although this system appeared secure and stable, grounded in biological imperatives and validating centuries of English dog breeding, in fact it resulted from an impressive collective act of will and imagination. Even the basic categories of the dog fancy—the breeds—were relatively recent constructions. The human upper crust might have been fondling lap dogs and following hounds for centuries, but there was little evidence of the relationship between these ancient animals and the highly bred and well-documented pets of the late nineteenth century. Indeed, the very notion of breed as it was understood by Victorian dog fanciers—a subspecies or race with definable physical characteristics that would reliably reproduce itself if its members were crossed only with each other—may have been of relatively recent origin. Dr. Caius's list of Tudor dog types bore little resemblance to modern schemes of classification. He recognized only seventeen varieties, far fewer than existed in the nineteenth century: "Terrar, Harier, Bloudhound, Gasehunde, Grehound, Leuiner, Tumber, Stealer, Setter, Water

The alternative to pedigreed dogs: a street vendor hawking mongrels
From Henry Mayhew, *London Labour and the London Poor*, 1861–62

PRESTIGE AND PEDIGREE

Spaniel or Fynder, Land Spaniel, Spaniel-gentle or Comforter, Shepherd's Dog, Mastive or Bande-Dog, Wappe, Turnspit, Dancer."[31] Some of these names anticipated those of nineteenth-century breeds, but it is unlikely that they referred to identical or even closely related animals. Caius's classification was based on function rather than physical appearance. He grouped his types under three larger categories, all unmistakably utilitarian: hunting dogs, pet dogs (this category included only the spaniel-gentle), and dogs that did menial work. Any large dog would have been called a mastiff; any lapdog a spaniel-gentle. An eighteenth-century sporting encyclopedia complicated these traditional functional categories with overlapping divisions based on coat color, which suggested that a dog with red spots was "fiery, and hard to be managed," while a yellow one was "of a giddy nature, and impatient."[32]

It was not until the eighteenth century that breeds in the modern sense began to emerge, starting with foxhounds, the most carefully bred dogs at the end of the century, although among the most "mongrelly" at its outset. Consensus about the desirability of the remodeled type emerged among Masters of Fox Hounds without institutional guidance or corroboration. By the early nineteenth century most hunts had adopted either the type embodied in the Quorn pack or the Brocklesby strain developed by the first Lord Yarborough.[33] When the 1795 records of the foxhound kennel at Beachborough classified individual dogs by breed, they identified the kennel from which the animals had been purchased. This informal reliance on a trusted strain persisted among some upper-class dog fanciers well into the nineteenth century, after the system of registered pedigrees had become firmly established. Thus in 1848 James Brierly of Mossley Hall, was able to determine that the earl of Derby's bloodhound bitch was that "rara avis," a real thoroughbred (hard to find because the few proprietors of such animals adhered to "principles of keeping the breed to themselves"), as a result of which he proposed mating her with one of his males of similarly exclusive descent. Aristocratic pug owners, in particular, preferred exclusive "private pedigrees" to those that had been publicly registered and were therefore available to anyone with sufficient money to buy them.[34]

These secluded and somewhat idiosyncratic records were not much use to the mid-Victorian regularizers of the dog fancy. And the only preexisting foundations for their taxonomical labors referred to sport-

Dogs classified by both breed and function. From R. W. Dickson, *A Complete System of Improved Live Stock and Cattle Management*, 1824

PRESTIGE AND PEDIGREE

ing breeds that never gained popularity as show animals. Foxhound pedigrees were routinely if unofficially maintained by the end of the eighteenth century. The next breed to attract genealogical attention was the greyhound. Early in the nineteenth century Major Topham of the Guards, a celebrated wit and sportsman, began to publish a register of matings and births among his coursers. But for most breeds the first *Kennel Club Stud Book*, published in 1874, initiated formal pedigree keeping, and even the entries that included some ancestry did not reach far into the past. One cocker spaniel's descent was traced back to 1866; a mastiff listed forebears for eight generations.[35]

The paucity of historical roots may have made fanciers more eager to specify the grounds of contemporary distinction that separated their dogs from the run of the canine mill, especially since the conventional boundaries not only between breeds but between the dog and similar animals turned out to be rather uncertain. Before the family history and personal attributes of individual animals could be certified as impeccable, it was necessary to assign them to a clearly defined category. This way it could at least be determined that their relatively recent family history and their personal attributes were beyond reproach. Thus the number of recognized dog breeds expanded steadily through the nineteenth century. In 1800 Sydenham Edwards listed only fifteen, although he restricted his attention "to what are termed the permanent . . . races," excluding the mixtures, "which, by repeated crossing in various breeds become almost infinite." Near mid-century, veterinarian William Youatt described almost forty varieties of the domestic dog, in addition to several wild breeds, but one-quarter of them were sub-breeds of greyhounds; a contemporary sporting encyclopedia listed over sixty breeds, including, rather confusingly, the "Prairie Dog of North America." Another index of category problems was class 220 at the Cruft's show of 1890, which consisted of "stuffed dogs, or dogs made of wood, china, etc."[36]

The profusion of new varieties led to recurrent attempts to systematize, in order, at least, to allow fanciers to make sense of the welter of show classes. The sporting encyclopedia offered six general, if motley, categories—wild dogs, greyhounds, hounds, fowling dogs, pastoral or domestic dogs, and mastiffs and their kind, a zoologist who cataloged the dog collection at the British Museum (Natural History) suggested a more sensible but equally unwieldy set of affinity groups based on

Prize Pets

wolves, greyhounds, spaniels, hounds, mastiffs, and terriers. The scheme Rawdon Lee used for his compendious, three-volume *History and Description of the Modern Dogs of Great Britain and Ireland*, which also recognized approximately sixty breeds, simply divided the animals into sporting and nonsporting categories. It was published in 1894, and the Kennel Club quickly adopted it as the official basis for classes in shows and in the *Stud Book*.[37]

All these lists and categories and family trees came to life at the shows that regulated and defined the dog fancy. Shows vividly illustrated the enormous range of available breeds as well as "the purest type" within each of them. In addition, by imposing complex controls on both animals and owners, they reenacted the elaborate structure of the fancy. Each dog was precisely defined by the class it competed in, which could reflect age, sex, color, and previously demonstrated merit, as well as breed. Dogs that won prizes in so-called championship classes were entitled to prefix their names with the coveted honorific "Champion."[38] When not in the ring, animals were (at least in principle) confined in assigned cages that were arranged in long rows and labeled to correspond with the program listings. During the judging process, decorum was expected of both dog and owner. More broadly, the same standards were extended to the crowds, and the transparent respectability of these highly regimented and ritualized occasions attracted new fanciers and enhanced the prestige of fancying as an avocation. The advent of modern dog shows purged dog fancying of an earlier, less savory reputation as suitable only for country squires, who needed foxhounds and shooting dogs, and for the rough urban types who liked to bet on bulldog matches and greyhound races. Show prizes were awarded for elegance and breeding, rather than brute strength and speed. The former were rubrics under which the self-consciously genteel—even ladies—did not hesitate to compete.

The rapidity with which the newly institutionalized dog fancy blossomed in the second half of the nineteenth century suggests that some such outlet for competition and demonstration was urgently desired. The first dog show was held in Newcastle on June 28, 1859. It was sponsored by a local sporting-gun maker named Mr. Pape, and there were only sixty entries in two classes (both for gun dogs: pointers and setters). Mr. Pape provided the prizes from his inventory. This seemed like such a good idea that a Mr. Brailsford organized a larger show of sporting dogs in Birmingham a few months later, which was, in turn, so successful with both dog owners and the public that it was repeated the next year, with the addition of thirteen classes for nonsporting dogs.[39] The inclusion of pet animals significantly increased the shows' popularity. By 1865 they had prospered so much that the organizers formed a company and built the Curzon Hall, where they continued to flourish. The first really large show—with over one thousand entries—was held in Chelsea in March 1863, and the first international dog show took place, also in London, two months later. Shows on the grand scale spread to Scotland in 1871, the date of both the first Scottish national show (held in Glasgow) and the first Scottish metropolitan exhibition (held in Edinburgh). Little more than a decade after their birth, dog shows had arrived at vigorous maturity.[40]

In 1900 one expert estimated that "taking out Saturdays and Sundays, there is a Dog Show being held somewhere or other on every ordinary day of the year." Already by 1889 there were fears that "we have *too many* shows," but the numbers continued to increase. There were 217 shows in 1892, 257 in 1895, 307 in 1897, and 380 in 1899. Most of these were small-town or regional affairs, offering local fanciers the chance to show off their pets and perhaps prepare for the big time. There was also a well-established national circuit of major exhibitions. Its high points were the Kennel Club shows (usually two a year—one in the summer and one in the winter) and the Cruft's show, all in London, and the National Dog Show, which took place in Birmingham.[41] The size of the major shows was another indication of the appeal of dog breeding. In 1890 almost fifteen hundred dogs competed at the Kennel Club show, and over seventeen hundred were entered in a show at the Crystal Palace. Provincial shows could be nearly as attractive: shows at Manchester and Liverpool the same year displayed over fourteen hundred competitors, and one at Brighton drew over one thousand entries.[42]

The rhetoric surrounding dog shows indicated not only their popularity, but also the recurrent concern of owners and breeders about the reputation of the dog fancy. Thus the shows sorted people, as well as dogs. In 1900 Charles Lane looked back "many years since" to the shows held in Warwick under the active patronage of the late earl of Warwick. These "delightful gatherings" were "admirably managed by a . . . courteous committee of 'real workers,' whom it was always a pleasure to meet"; they attracted "the 'Flower of the Fancy,' both dogs and people."

The status of dog breeding was an issue, apparently, that could not simply be resolved and forgotten. Much closer to the time when Lane was reflecting complacently on the past, fanciers were proudly pointing to recent triumphs over even the shadow of vulgarity. In 1894 the *Ladies' Kennel Journal* announced that one of the objectives of the organization that sponsored it, the Ladies' Kennel Association, was "to prove that, though [its members] may be . . . what a certain class of critic is pleased to call 'doggy' women. they are, none the less, gentlewomen." No less averse to tooting its own horn in this connection, the Kennel Club announced several years later in its journal that "the efforts of the Kennel Club have resulted in elevating the general tone of canine exhibitions."[43]

This at least was the canonical picture, promoted by the Kennel Club and other guardians of the carefully constructed establishment. They recognized, of course, that Mr. Pape had not conceived his dog show in a flash of blinding inspiration. He was implementing a suggestion made repeatedly by thoughtful dog breeders and tried unsuccessfully at least three times before 1859.[44] In 1847 Robert Vyner, a one-time Master of Fox Hounds, suggested that "nothing would be more likely to improve the breed of fox-hounds than prizes, to be awarded by competent judges." In 1854 "Bow-wow," a correspondent of the *Courser's Annual Remembrancer, and Stud Book*, argued that, since "the greyhound is a purely national animal, brought to greater perfection in this country than any other, and it is unrivaled for its symmetry, beauty, and points of breeding . . . why should we not have an 'Exhibition of Dogs'"—an exhibition which, he hoped, would also include mastiffs, deerhounds, and bloodhounds.[45] Both these fanciers based their suggestions on the readily available example of livestock shows, which many of the gentry breeders of sporting dogs attended regularly; indeed, some of the dog shows that were subsequently established, including that held in Birmingham, owed much of their box office success to being scheduled at the same time as important agricultural exhibitions.

If the officially promulgated line of descent for dog shows derived them from well-established livestock shows, there was also a somewhat shadier collateral lineage. Dog shows of a sort had been going on long before 1859, although they lacked elaborate class divisions and certified judges. Public houses were their most frequent venue. Hugh Dalziel, an eloquent and distinguished breeder of collies under the new dispensation, remembered the "convivial meetings" of what he called "the pre-historic age of dog shows" (when his own main interest had been bulldogs) rather fondly. Every member of the company was exhibitor, spectator, and judge; they examined each other's animals, then arrived at a noisy consensus. The gatherings were often subsidized by landlords to attract customers with "doggy proclivities," who were apparently considered desirable additions to a tavern's clientele. In the 1840s Jemmy Shaw's establishment near the Haymarket and Charley Aistrop's in St. Giles were favorite London locations for these one-breed shows; they were held in rooms reserved for rat-killing competitions on other evenings.[46] Sporting and fighting breeds were featured most frequently, but even aristocratic pet breeds might be displayed in these inelegant surroundings. Lytton recorded a toy spaniel show as early as 1834 at the Elephant and Castle, and one in 1851 at the Eight Bells. In acknowledging such precedents, as in other ways, she outstripped most of her fellow fanciers in social candor. According to Dalziel, in "polite circles" the birth of dog shows in "sanded parlours" was "generally ignored"— one of many instances when "it is felt to be inconvenient ... to trace the pedigree too curiously, lest the low origin might be found inconsistent with existing pride."[47]

The behind-the-scenes realities of dog shows often belied their superficial order and decorum, however, paradoxically, one index of how much fanciers valued the hierarchical ideal embodied in the judging process was the amount of divergence from it they were prepared, albeit unwillingly, to tolerate. If badly designed, the neat accommodations prepared for competitors could quickly become chaotic. At the 1863 Chelsea show, according to Stonehenge, the lavishly decorated hall seriously misrepresented conditions below. No water or water dishes had been provided for the animals; worse, the dogs were inadequately caged and on chains so short that "their owners were continually letting them out." One elderly lady was bitten in the arm by a retriever.[48]

These problems threatened the shows' participants as well as their image. In exchange for the chance of a prize, owners had to risk their cherished pets' lives. Even shows offering reasonably comfortable and sanitary accommodations frequently spread distemper, which was particularly dangerous to competitors in the puppy classes. And if an animal survived the show itself (where the behavior of spectators might present additional hazards—at a cat show held in Birmingham in 1875, one admirer held the lighted end of his cigar to the nose of a prize-winning tom), it had still to brave the journey home.[49] The railroads' treatment of animals was a constant subject of complaint by humane organizations, as well as dog fanciers. Dogs were allowed to suffer, and sometimes to perish, from cold and hunger; they were often lost, and at the best of times they were transported "in a dirty den called a boot—rank with accumulated filth of every kind."[50]

It was bad enough when the incompetence of show organizers endangered the health of the competitors. Much more outrageous, in the view of fanciers, were careless administrative arrangements that jeopardized the whole purpose of the show. In a second and severer assault on the 1863 show, Stonehenge detailed further horrors. Dogs were judged with the names of their owners attached to their collars, instead of the numbered tickets that would have preserved anonymity. The arrangement of the exhibits repeatedly violated the integrity of classes: retrievers were mixed up with setters; foreign dogs were lumped together indiscriminately; and a group of puppies occupied the bench allotted in the catalog to "Wolf, brought from the Crimea." In a show that featured confusion, rather than clarity, order, and rank, it was not surprising that some observers (although not Stonehenge) questioned the probity of the judges.[51]

From their initial appearance in 1859, public dog shows had been dedicated to the achievement of three related goals: to improve the various breeds of dogs, to display model specimens, and to discourage the breeding of mongrels. When they selected the prizewinner in each class, therefore, judges were not simply recognizing a particular outstanding animal. At the same time, they were identifying the strain to which the prizewinner belonged as promising breeding material, and they were endorsing a type toward which other breeders should aspire. Especially in awarding the prizes at important shows, which were widely reported and criticized in the press, judges wielded considerable power. If they were not perceived to exercise it reliably and responsibly, the elaborate structure of dog showing, which rested at bottom on public confidence in their honesty and discrimination, would collapse.

Yet, especially in the first two decades, it was hard for even the most competent and conscientious judge to do his or her job properly. There was, to begin with, too much to do. Large shows might have only ten judges for several hundred classes.[52] In addition to assessing the quality

PRESTIGE AND PEDIGREE

of the dogs on display, a judge had to be alert for "faking"—practices one dog-showing manual summarized as "plucking out white hairs . . . , dying the coat, cutting the cartilage of the ear to make it droop, shortening a tail that is too long . . . , cutting the coat with scissors." Nor did the exhibitors make the job easier. In the ring, fanciers frequently tried to maneuver their animals so as to cut off the judge's view of other competitors; they were also apt to offer distracting praise of their own dogs and criticism of rivals.[53] This was all in the spirit of good clean fun, the kind of thing the most refined breeder might try in order to gain coveted recognition for his or her pet.

Behind the scenes there was actual fraud. Animals were doctored, misleading or false descriptions were inserted in catalogs, buyers who had paid for the stud services of champions, or for the champions themselves, were swindled with inferior animals. Even the lesser animals ordinarily purchased by new fanciers might be similarly misrepresented; Gordon Stables recommended that inexpert purchasers avail themselves of the approval service offered by the editorial offices of several periodicals, which would hold their money until the animal and its pedigree had been certified genuine and acceptable.[54] The people who perpetrated such crimes were dangerous to the dog fancy on two counts. They cheated individuals, and their shady reputation might impugn the respectability of the fancy as a whole. For none of those who considered themselves disinterested dog lovers—whose motives in showing, that is, were untainted by the desire to make a profit—doubted for a moment that these criminals were "pure dealers . . . they show for the purpose of winning the money prizes, and to enable them to sell the dogs they breed." They were unsavory holdovers from the bad old days of tavern shows.[55]

To preserve the newly genteel reputation of dog shows, a group describing its members as "true sportsmen . . . who breed to win and to whom pecuniary questions are of no moment" formed the Kennel Club in 1873. Its main initial concern was to combat fraud by establishing the identity and descent of pedigreed dogs. At the same time it developed an interlocking system of shows and registration designed to limit competition to a carefully screened segment of the canine, and, implicitly, of the human population. The first volume of its *Stud Book*, which listed dogs exhibited since 1859, appeared in 1874. Geared to the show

Intense competition for prizes in the show ring. From Judith Neville Lytton, *Toy Dogs*, 1911

PRESTIGE AND PEDIGREE

circuit, the *Stud Book* concentrated on dogs that had won prizes. Its editor acknowledged that some "excellent and well-bred dogs" might have been excluded as a result of this policy; but, he argued somewhat circularly, "the value of the blood and pedigree is demonstrated on every page, as nine-tenths of the later prize winners trace back to prize blood." Keyed to the *Stud Book* was a national registration system for pedigreed dogs The *Stud Book* also included a code of rules for dog shows, one of which prohibited any unregistered dog from entering a show held under the auspices of the Kennel Club.[56]

The Kennel Club encountered resistance in implementing these reforms, sometimes from the disreputable operators whose business interests were threatened, sometimes from competing groups like the organizers of the Birmingham show, who resented what they viewed as a national power grab. But most fanciers strongly preferred dignity, discipline, and quality control to the *caveat emptor* ethos of the untrammeled market. So the Kennel Club triumphed relatively quickly. By 1892 its fighting days were far behind it, in the view of a rising generation of well-bred fanciers.[57] But in a way this evidenced the thoroughness of its achievement. Under its guidance dog showing had left its raffish origins behind. It had become a respectable and well-regulated pastime, a reflection of the carefully calibrated human social order.

Determining Breed Standards

The structures that evolved in the third quarter of the nineteenth century to regulate the breeding and showing of pedigreed dogs figuratively expressed the desire of predominantly middle-class fanciers for a relatively prestigious and readily identifiable position within a stable, hierarchical society. Most of the institutions of the fancy—shows, stud books, and breed societies—had been borrowed from high stock breeding, and their conservative symbolism survived the translation. Yet the Kennel Club was not simply a scaled-down imitation of the Smithfield Club. While recognizing the seductive charm of aristocracy, the Victorian dog fancy embodied a set of values that undermined the traditional social code that structured high stock breeding. The identification of elite animal with elite owner was not a confirmation of the owner's status but a way of redefining it. Thus a metaphor that signified stability in the world of prize cattle signified change in the world of prize dogs And although pedigree was important, the rhetoric of distinguished descent was invariably strongly modified by a more individualistic ethos. Dog owners were more intimately involved in the web of competition than the owners of prize livestock; they even appeared in the ring along with their animals. Breeds might flourish as a result of elite associations, but inherited rank did not ensure a distinguished career for an individual animal. Although a respectable pedigree was essential as a "guarantee of purity of blood," it did not guarantee anything beyond that. Indeed, experts warned that the inflated pecuniary value of the offspring of celebrated champions could result in the preservation of inferior animals in the most distinguished pedigrees.[58] Becoming a champion was primarily an achievement of the individual dog and owner, and only secondarily a ratification of the animal's ancestry.

In fact, the dog fancying establishment was structured by two apparently complementary but ultimately inconsistent metaphorical systems. The uneasiness it inspired among aristocrats like Lytton and sporting country squires reflected more than a desire to defend their turf from interlopers. At the same time that the elaborate divisions of dogs into breeds and classes, and of individuals into precisely ranked hierarchies within those classes, seemed to imitate and thus endorse the stable, rigidly hierarchical social system represented by the human upper orders, the content of those discriminations—or rather, their lack of content—undermined it. As Lytton sensed, the problem was what was ranked. The hierarchy projected through the display of prize livestock consisted of ranks determined by concrete and stable differences among people. Similarly, the dog breeds most closely associated with gentry fanciers—greyhounds, foxhounds, and shooting dogs—were judged not in the artificial environment of shows but by their performance in the field. Among show pets, however, rank was divorced from any practical criterion, many breeds were judged according to standards set simply for the sake of making distinctions. In 1866 Stonehenge charged that "at present, exhibitors do not know what points will ensure a prize, and all is confusion." Several years later he complained that, although dog shows had multiplied and increased in popularity, there was still "no standard of merit by which to be guided in breeding and judging. In many cases the choice of points is wholly arbitrary, as in the toy dogs; for no one can contend that on any other principle a small eye shall be a merit in one breed (toy terrier) and a defect in another (King Charles Spaniel)." Sometimes qualities were valued only because they were unusual or difficult to produce. The prizewinning pedigreed dogs of the late nineteenth century seemed to symbolize simply the power to manipulate and the power to purchase—they were ultimately destablizing emblems of status and rank as pure commodities.[59]

Even if there was no reason to prefer one set of points over another, fanciers agreed that each breed required a model or standard so that the central business of classification and ranking could proceed. Most of the breeds judged at late nineteenth-century dog shows had therefore been created out of whole cloth, or something very close, in the not-too-distant past. The rather ferocious selection pressures that shaped the evolution of the foxhound did not provide equally clear direction even for all sporting dogs. The beagle, for example, had "a place among the very oldest English breeds," as one aficionado proudly claimed in 1898. Rumor identified it with the "brach" of past generations, and legend suggested that good Queen Bess had owned a pack of the animals small enough to fit in her glove. Yet no one was quite certain what a beagle was like. It was certainly a hound, and probably the smallest kind of hound. How small was one of many vexed questions. In 1879 Vero Shaw, a vigorous and intelligent observer, had suggested that the difference between contemporary beagle packs was "certainly chiefly one of size." The author of the *Kennel Gazette's* "Retrospect for 1898" reported happily that "we are gradually getting back to the little compact hound," and in the same year Shaw warned prospective beagle buyers not to "pay much attention to an animal over 14 in. high at the shoulder"—although a critic hoped that "so old-fashioned a breed" would not be "constitutionally ruined to satisfy a modern rage for undersized specimens." Beyond this there was less agreement. One reporter complained, after the 1888 Kennel Club show, that "there are so many sizes and types of beagles that it must always be a very unsatisfactory task judging them when only one class is provided."[60]

Yet all the disputants assumed that a standard or ideal beagle existed, even though there was no historical evidence for a single, true, ancient beagle type. The heyday of the beagle had been in the slow, unfashionable days of hunting, before sportsmen became particular about the species of their quarry or the lineage of their dogs. Already in 1800 Edwards observed that beagles "are by no means in such repute as formerly, a complete pack of them being rarely seen." Dashing sportsmen preferred stronger, faster dogs, and were repelled by the beagle's fondness for chasing the lowly hare. A beagle enthusiast noted wishfully in 1823 that although beagles had been "almost abandoned, owing to the introduction of the fleeter and more fashionable harrier . . . more packs are now to be met with than . . . a few years since," which he attributed to the fact that their slow pace made them easy for timid riders, ladies, and unmounted rustics to follow.[61] He may have been premature in his celebrations, however. In 1861 there were still only a few beagle packs, and at the Kennel Club show of 1894, the beagle classes were canceled because of insufficient entries.[62]

This spotty record gave the Beagle Club little basis on which to begin its work when it was formed in 1895 to further "the true interests of the breed."[63] The founders of the Beagle Club admitted that "the true Beagle is in great danger of disappearing owing to cross and careless breeding," and they recognized that the difficulty of their task was compounded because "there is no standard to breed to, and . . . no two Beagle judges are of the same opinion." But the club took its mandate seriously. Where there was a name, there was a breed, and where there was a breed, there had to be a standard. The prestige of the club's members was at stake along with that of the animals with which they identified. They set themselves a double task: "to improve the make and look of the Beagle; and at the same time to render it more suitable for work." In a sense the club succeeded. Whether the breed was restored and improved, or newly imagined and created, enforceable standards were developed. And once ranking became predictable, beagles began to increase steadily in popularity and prestige.[64]

Like the beagle, the bulldog ostensibly sported a long history. Although Caius did not include it among his British dogs, the bulldog's advocates considered it coeval with and emblematic of Anglo-Saxon civilization. Pierce Egan, a chronicler of the early nineteenth-century sporting scene, had characterized it with the thoroughbred horse, the game cock, and the pugilist as one of "the peculiar productions of BRITAIN and IRELAND, unequalled for high courage, stoutness of heart, and patience under suffering."[65] And to an even greater extent than the beagle, it had survived its original function and been redesigned according to standards that seemed completely arbitrary. In 1900 it was "amongst the most popular breeds" of pedigreed dogs, pampered, according to one fancier, more than any other variety; another identified bulldog puppies as particularly "delicate." Bulldogs were said to be so indolent that they did not have much natural appetite, but "as a *thin* Bull Dog is an abomination," it was necessary to coax them to eat. Although fanciers always liked to see their special breed rise in public favor, the bulldog's wide acceptance caused a kind of identity crisis among its most faithful adherents. A show of old-fashioned bulldogs would not have attracted a more "fashionable crowd" than other dog shows; no one had ever referred to the traditional article as "a ladies' dog, as its kindliness of disposition admirably fits it."[66]

The good old days of bulldog fancying ended in 1835 when Parliament made bull baiting illegal. Before then bulldogs, like foxhounds, had been bred for function; indeed the category may have identified not a breed but a motley group of similarly talented animals. The requirements of bull-to-bulldog combat were exacting. Enraged bulls charged with their heads down. In order to avoid the lethal horns, the dogs had to be low to the ground and relatively nimble. Because a bull was most vulnerable on its tender nose, the bulldog needed strong jaws as well as the dumb courage to jump at the bull's face and the perseverance to hang on. The quality of the animals was tested regularly under fire, and prowess was more important than appearance in determining rank. Late eighteenth- and early nineteenth-century portraits of renowned bulldogs showed animals that varied widely in size and shape, even making allowance for the unequal skill of the artists. Some even lacked the characteristic bulldog countenance, retrospectively described as one in which "the broad, projecting underjaw ensures the terrible tenacity of grip; the wide nostrils, placed far back, enable the dog to draw unimpeded breath while keeping his teeth fixed on the yielding cartilage of the bull's nose."[67]

After the abolition of bull baiting, the bulldog went into a decline. Within twenty years it was announced that "this fine specimen . . . is at the present day almost dying out." When the Bulldog Club was launched to save the breed in 1874, the bulldog was "at a discount." The unsavory surroundings with which the bulldog had been associated may have had almost as much to do with its near extinction as the ban on bull baiting. Although it might occasionally have been the pet of ladies and gentlemen (Lady Castlereagh, the wife of the Regency foreign minister, was reputed to have driven in the park with a large bulldog on the

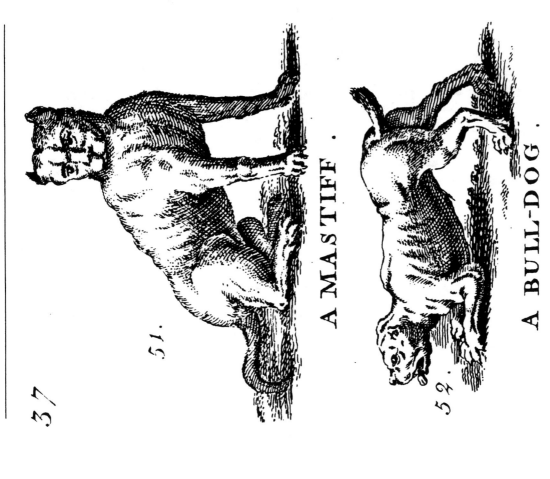

A mastiff and a bulldog, breeds hard to distinguish in the eighteenth century. From Thomas Boreman, *A Description of Three Hundred Animals*, 1736

seat beside her), its patrons were, as a rule, not among "the better class of fanciers"; and the places where specimens were generally to be found earned the breed the name of "the pot-house dog."[68] William Youatt asserted that although keeping such sporting dogs as spaniels, pointers, and even greyhounds caused "no diminution in respectability," a young man with a bulldog would "speedily become profligate and debased." In addition, the "fondness of the lower orders in some districts for the

A pomeranian with an early nineteenth-century bulldog, shown lacking many of the characteristics later associated with the breed
From John Church, *A Cabinet of Quadrupeds*, 1805

fighting and baiting propensities" of bulldogs continued after 1835, and there were doubtless still opportunities for the animals to show their stuff. Those dogs not kept for fighting were "principally bred by professional dog-fanciers," the lowest echelon of fancying society.[69]

So the fledgling Bulldog Club had its work cut out for it: to redefine a breed that had outlived its usefulness, that had no social cachet, and that appeared to ordinary dog lovers ugly, stupid, and brutal. It was, in the words of the club's official description, "much maligned and . . . very little understood."[70] The first step in the rehabilitation of the bulldog was rhetorical: to blame people for the unpleasant qualities that had been attributed to the dog. While Youatt had characterized the bulldog as "scarcely capable of any education, and . . . fitted for nothing but ferocity and combat," and the lion tamer Van Amburgh had considered it as brave as any wild beast but "rather deficient in its range of ideas," later writers pointed to extenuating circumstances. According to Dalziel, the bulldog's courage made it "the only dog with sufficient endurance to serve the cruel purposes of depraved owners." And the strategy worked, the newly imagined bulldog caught on among respectable fanciers. By 1885 it was second only to the collie in popularity, as measured by dog show entries. It began to be described as "peaceable" and "intelligent," even "benign-looking." An advertisement for the sale of a champion named Bully McKrankie noted that "he has always been kept in the house and is a great pet." Bulldog shows drew larger gates than any other one-breed show.[71]

As soon as the breed took its place in the show ring, however, it became clear that no generally accepted or rationally founded breeding standards had replaced the old law of survival of the fittest. A correspondent who had "only quite recently entered the Bulldog Fancy" implored the editors of the *Sportsman's Journal and Fancier's Guide* to favor "green fanciers" with a brief description of "the points, general make and shape . . . of the bulldog." His independent research had, he complained, only compounded his confusion: "At present . . . after trying to collect some information, we are worse off than when we commenced." Almost any feature of the animal was open to debate. Veteran aficionados offered contradictory opinions, and, true to the bravura spirit of the bulldog fancy, they offered them with ringing confidence.[72]

The Dudley nose question, for example, convulsed the Bulldog Club for over a decade. (Dudley or flesh-colored noses occurred in some

The new model bulldog of the Bulldog Club
From the *Kennel Gazette*, 1899

strains of fawn-colored bulldogs, usually in conjunction with light eyes and a yellowish countenance.) In 1884 the club voted to exclude dogs with Dudley noses from competition, defeating a counterproposal that they be considered *sine qua non* in fawn bulldogs. Dudley nose advocates lobbied energetically among club members; a spirited correspondence aired in the canine press. But the majority repeatedly reaffirmed its position, although in 1895 it conceded that dogs with part-Dudley noses (black and flesh-colored) could "compete and win if they were sufficiently good in other respects." Experts outside the polarized club membership viewed the issue with greater detachment. "Personally," confessed Rawdon Lee, "where a dog is otherwise good, I would not disqualify him for Dudley markings."[73] More functional features were also subject to the same kind of arbitrary assessment, sometimes in defiance of the animal's practical requirements. One *Kennel Gazette* reviewer claimed that to be "distingué" a bulldog should be "well out at the shoulder and with a good broken up face"; a dog lacking either was doomed to mediocrity. But Lee characterized this shoulder configuration as a crippling deformity. He cited the example of Dockleaf, a renowned champion of the early 1890s and the property of the eminent bulldog breeder Sam Woodiwiss, which collapsed after walking a mere two miles.[74]

Although experts agreed that the late Victorian bulldog differed noticeably from its fighting ancestors, they disagreed in their assessment of the new model. Stonehenge thought that "the modern bulldog" was "decidedly neater in shape" than its forebears, but a few traditionalists regretted the change. Some characterized it as simple declension—modern bulldogs were less active, and "want of character" was prevalent among them. Others were harsher in their judgments. According to George Jesse, a self-appointed defender of oppressed canines, "the disgusting abortions exhibited at the shows [were] deformities from foot to muzzle ... and totally incapable of coping with a veteran bull." There were, however, no more bulls to cope with. The only influence on the make of the bulldog was the pleasure of its admirers.[75]

The collie was the most popular pet dog in late Victorian England and a prime example of a breed reconstructed to express the figurative needs of fanciers. Collies were originally valued for the qualities they had developed as hardworking Scottish sheepdogs—intelligence, loyalty, and a warm shaggy coat. Once they were firmly established in the *Stud Book*, however, breeders began to introduce modifications and improvements, which were tested not against the rigors of the Highland winter, but in the fashionable marketplace. By 1895 there were seven independent clubs devoted to the breed's welfare, many of which sponsored all-collie shows, as well as strong collie representation in the Kennel Club and regional canine associations.[76] The large number of pedigreed collies seems to have exacerbated the tendency of fanciers to fabricate subtle points of distinction between animals and artificial models to measure them against. As a result, fashions changed swiftly and collie standards were among the most volatile, breeders redesigned their animals or restocked their kennels in accordance with the latest show results. Plasticity could even take precedence over pedigree; in order to instill some temporarily admired attribute, breeders were sometimes willing to contaminate the strain. In the early days of showing, collies were frequently crossed with Gordon setters, to achieve then-fashionable glossy, black-and-tan coats. For decades experts could detect "traces of this bar sinister"—the telltale ears, head, and general heaviness—in many show animals. Even without crossing (which became less common as the *Stud Book* gained sway), fashion could undermine the character of the breed. The 1890s saw a craze for exaggerated heads with long, pointy noses. In 1891 a *Kennel Gazette* reviewer com-

DOG FASHIONS FOR 1889.

Extravagances of fancy dog breeding. From *Punch*, 1889

plained that show judges had given all the prizes to "dogs of this greyhound type whose faces bore an inane, expressionless look." Critics alleged that such dogs could hardly display the intelligence characteristic of their breed because there was no room in their heads for brains.[77]

Specialist clubs were supposed to defend their chosen breeds against the vicissitudes of fashion, but they had few other guides in their attempts to establish standards for breeders and judges. Their eagerness to establish the separate identity of their chosen breed inclined club activists to focus on "arbitrary and conventional points" that were easy to define, and to neglect the character and bearing of the whole animal. The Collie Club was no exception. It was even proposed, at a Collie Club meeting in 1884, that collie classes at shows be reorganized according to color. This notion, however, went too far and was overwhelmingly rejected by the membership. Such distinctions were not sufficiently subtle or difficult to produce credible rankings. In addition, adopting them would implicitly have compromised the gentility of the collie fancy by linking it to the color-coded displays of fancy mice, rabbits, and guinea pigs that were recognized as the preserve of the vulgar.[78] From this point of view the diminished utility of working breeds might seem at-

tractive to fanciers, as a barrier between the dogs' original breeders and their new patrons. There was little overlap between the rough shepherds who bred collies for work and the affluent amateurs who bred them for show, and fanciers were inclined to believe that the dogs had benefited from their transplantation. Although the original working collie was apt to be rude and hostile to strangers (and likely to savage the sheep in neighboring flocks when off duty), "constant association with his superiors has improved his disposition immensely." The dog had become worthy of its elegant new surroundings, "despite his lowly origin (for, after all, he is but a sheep dog)."[79]

On the whole, however, newly codified breed standards were designed to privilege rank by competition rather than rank by association. The juxtaposition of arbitrarily established criteria (the major purpose of which was to make judgment possible) with swiftly changing fashions not only in favorite breeds but in preferred types within those breeds symbolized a society where status could reflect individual accomplishments and was, as a result, evanescent, lacking in foundation, and in constant need of reaffirmation. As most dog fanciers were, in this sense, self-created, so their exploitation of the physical malleability of their animals was extremely self-referential. Its goal was to celebrate their desire and ability to manipulate, rather than to produce animals that could be measured by such extrinsic standards as utility, beauty, or vigor. Thus it was an index of their paradoxical willingness aggressively to reconceive and refashion the social order in which they coveted a stable place.

The Cat Fancy

If the elaborately distinguished and carefully graduated classes of the dog fancy mirrored breeders' desire to improve their own social positions, the subsequent development of a different animal fancy highlighted the inconsistency inherent in the rhetoric of pedigreed pet breeding. *Canis familiaris*, the domesticated dog, is among the most plastic of species. The enormous variation in size, shape, temperament, and athletic ability among individuals and breeds might have seemed to explain—and, indeed, to have required—the complex subdivisions that quickly came to characterize dog shows. The reverse is true of *Felis domesticus*, the domesticated cat. Members of this species differ from one

PRESTIGE AND PEDIGREE

another in little but their coats. Indeed, in the early years of dog showing, cats were not prestigious fancy animals. If they were shown at all, it was usually as an addendum to an exhibit of rabbits or guinea pigs, which were also distinguished by such superficial characteristics as fur length and color.[80] But in their relations to their owners, pet cats were much more like dogs than like the rabbits, rodents, and prize poultry that were widely bred for purely ornamental purposes, and their owners were almost as likely as dog owners to identify with them. As a result, cat advocates constructed an analogous series of institutions to classify and evaluate their pets.[81]

Like the pioneering dog fanciers of mid-century, the architects of the organized cat fancy decided that their first task was to construct a taxonomy. But before they could establish a system of breeds and pedigrees, they had to solve a problem that their predecessors had not had to confront. Cats' mating habits made it difficult for fanciers to influence the appearance of the next generation, or even to know its parentage. Unlike dog breeders (and livestock breeders, for that matter), whose animals were usually willing to mate with a designated partner, cat breeders had to cope with animals accustomed to making their own decisions and implementing them out of sight of their owners. As late as 1868 Darwin claimed that people had done "nothing by methodical selection; and probably very little by unintentional selection," to influence the development of the domestic cat except saving the prettiest kittens in each litter and killing adults that poached game.[82] Thus the engagement of early cat fanciers in their animals' reproductive activities was more passive than that of dog breeders, and the physical evidence of their influence less compelling. Because of the difficulty of maintaining pure strains, there was no guarantee that a prize animal would produce prize offspring, and a sharp-eyed fancier might well spot a cat with magnificent show potential prospecting among the garbage. Consequently, the prices commanded by late Victorian prize cats did not approach those of dogs. As little as £1–£2 was regarded as a "long price" for kittens "good enough to win at a first-class exhibition."[83]

The inherent resistance of the species to subdivision and genetic manipulation did not discourage cat fanciers, who assaulted these biological impediments by establishing breed categories, each with ascribed standards and a hierarchy of animals within it. The first cat show took place in the Crystal Palace on July 16, 1871. It was organized by Har-

An early feline champion,
which was not too different from an ordinary cat
From Harrison Weir, *Our Cats*, 1889

rison Weir, a well-known writer about and illustrator of animal subjects, who later became president of the National Cat Club. His objective in proposing the show was appropriately taxonomical, although he characterized the endeavor in terms of discovery rather than invention: to encourage greater attention to "the different breeds, colours, markings, etc.," and he projected an elaborate structure of classes for "the different varieties of colour, form, size, and sex." As at early dog shows, despite these finely drawn subdivisions, there was some confusion about the boundary between species. One of the prizewinners was a Scottish wild cat. Cat owners proved as eager as dog fanciers a decade and a half earlier for a chance to display and compare their animals. Weir's show was so successful that the Crystal Palace Company, which had allowed him to perform his duties as organizer as a "labour of love of the feline race," rewarded him with a large silver tankard. Large cities like Birmingham and Glasgow immediately followed the metropolitan example, and within a decade or so even the smaller provincial towns could "boast of an annual exhibition of feline favorites."[84]

That the class categories proposed for these shows relied on color differences, the only readily available index of "breed" distinctions, revealed their weakness, although fancy organizers tried to disguise the problem by associating color with less superficial characteristics. One early attempt was made by Gordon Stables, active in the cat fancy as well as the dog fancy, who suggested thirteen categories: tortoiseshell,

tortoiseshell-and-white, brown tabby, blue and silver tabby, red, red tabby, red-and-white tabby, spotted tabby, black-and-white, black, white, unusual color, and any other variety. Stables was careful to cover himself against objections, explaining at one time that his purpose was "to describe the classification of cats generally adopted at pussy-shows, instead of dividing them, as they otherwise ought to be, into the different species and breeds" (although he did not specify these breeds). On another occasion he asserted that, although there was "properly speaking" only one breed of domestic cat, "colour is often the key to [its] character . . . , to its temper . . . , to its qualities as a hunter . . . , and its power of endurance." He claimed, for example, that tortoiseshells were "good mothers [and] game as bull terriers," while tortoiseshell-and-whites were "very clever, docile, and tricky"; black cats were "noble and gentlemanly," whites "far from brave . . . fond of . . . society . . . gentle . . . often delicate"; the black-and-white "sometimes . . . did not trouble himself too much about his duties as a house-cat."[85]

Stables' categories soon went out of fashion, largely because most cats shown in the 1880s and 1890s would have fallen into his last two classes. But what replaced them was not very different. The scheme that Weir developed for the Crystal Palace shows was also color-based, although it specified a broader range of colors and divided animals into shorthair and longhair classes as well as by sex and age. (The 1889 Crystal Palace show, which attracted 511 entries in 54 classes, grouped 102 of them under a division that was not part of Weir's original plan but that reflected the concern of dog fanciers that animal hierarchies reproduce human ones: Cats Belonging to Working Men.) Subsequent schemes were simply refinements. Eye color, for example, gradually became a more important component of judging, and categories were added to accommodate newly imported strains like the Siamese and the Abyssinian. The specialist cat clubs, which began to spring up at the turn of the century, also followed color lines; among the first founded were the Silver and Smoke Persian Cat Club and the Orange, Cream, Fawn and Tortoiseshell Society.[86] As the National Cat Club, modeled on the Kennel Club with its interlocking system of pedigree registration and closed shows, assumed control of the organized fancy, breeders began to claim that "a much greater study was made in the science of Cat-breeding."[87] But this had no noticeable effect on the category divisions, which, despite constant effort on the part of fanciers, could not

be made to imply much beyond color. Yet fanciers were never tempted to abandon a taxonomical enterprise that kept petering out in triviality; instead, they struggled the harder to justify complex class divisions by endowing them with greater significance.

The distinction between longhairs and shorthairs was more productive than those based on color, because the first represented exotic "oriental" imports (more or less interchangeably called Persians and angoras), while the second was the native English type. Stables identified shorthaired tabbies, in particular, as "the real old English cats—the playmates of our infant days and sharers of our oatmeal porridge." Nevertheless, he preferred "Asiatic cats" as pets. Though not so hardy as shorthairs, or so good as catching mice, they were "extremely affectionate and loving."[88] Subsequent fanciers agreed on the dichotomy while differing about its implications. One found longhairs "lazy and listless," though fashionable, and "not such pleasing companions" as the "sleek, agile, graceful, and intelligent animals with which we are more familiar"; another considered Persians "not so amiable" as shorthairs, but "more intelligent" and more inclined to "make themselves at home in their surroundings." Whatever their nature, fanciers clearly preferred foreign cats. Of 740 entries at the Crystal Palace show of 1896, 485 were longhairs, 64 foreign shorthairs, and 191 English shorthairs. "It almost looks," lamented a patriotic reporter, as though the native breed "were being dropped as a fancy animal."[89]

Many of the foreign shorthairs were Siamese cats, which became relatively popular in the mid-1890s. Previously they had been quite rare; barely twenty individuals were exhibited between 1871 and 1891, and as late as 1896, the *Ladies' Kennel Journal* could identify only four English breeders. Victorian legend held that the animals were the special property of the King of Siam, the few cats that made their way to Europe were considered either very special gifts or the booty of daring thieves. But "great changes in Siam" in the early 1890s (the result of pressure from British and French colonizers in neighboring territories) were believed to have produced a "comparative abundance of imported cats." The Siamese was praised for its beauty and temperament, and its romantic origin enhanced its appeal. But it is likely that among the Siamese's most powerful attractions was its very distinctness, which made it incontestably a breed apart.[90]

Even more than the dog fancy, the cat fancy celebrated the primacy

PRESTIGE AND PEDIGREE

of process over content. A typical turn-of-the-century claim about the progress of cat breeding distilled the circularity and self-reference of the enterprise: "Classes at shows on more extended lines have been established, and they are now judged by those who really understand the subject, therefore each variety is sure to improve; especially when it is clearly laid what has to be bred up to." The abstract idea of hierarchy was more important than the concrete basis of ranking. As standards were being developed, often the determining criterion was, as with dogs, simple rarity. Thus tortoiseshell toms were considered desirable because orange-colored cats are usually female—indeed, it was widely if incorrectly rumored that the Zoological Society of London had offered £250 for such an animal—and brown eyes were prescribed for black cats just because they seldom occurred. Yellow, the most common eye color, was analogously denigrated as an indication of low quality.[91] Breeding standards functioned to identify the animals best adapted for the purposes of the fancy; the most satisfactory criteria were those that allowed distinctions to be made and prizes awarded most efficiently.

Not all these distinctions were founded in biology; cat fanciers were relatively unsuccessful in reproducing among their animals the physical variety exemplified by dog breeds. This did not, however, compromise either the enthusiasm with which they prosecuted their objectives or the way in which they structured competition. Problems that resisted solution on the level of applied zoology could be attacked on the higher plane of rhetoric. The elaborate categories of the cat fancy were, among other things, an exercise in projection and fantasy; most feline breeds were verbal rather than biological constructions. Even among domestic animals, pets were especially available for such manipulation. They were utterly at their owners' disposal in more ways than one; they were chattels, and there was no significant human counterinterest to reinterpret fanciers' rhetoric or reconstruct their taxonomy. *Punch*'s intermittent jibes at the sentimentality of pet owners, and the criticism of their extravagance that occasionally appeared in other periodicals, hardly constituted significant challenges to the figurative structure of fancying. Even aristocratic stock breeders confronted more substantial rhetorical resistance, both from the technical experts who reinterpreted a shared set of images to support a divergent social ideal, and from consumers and humbler participants in the meat industry, who rejected both their values and their metaphors. Unlike prize cattle, pets were

Prize Pets

widely acknowledged to be useless, except for emotional and rhetorical purposes. This was, however, a function of their status rather than their species. At about the same time that pedigreed pet dogs were being marginalized as mere objects of their owners' indulgence and symbols of their aspirations, other less fortunate dogs became the focus of a serious confrontation between human groups. In this case realities of power generated a more anxious and confused rhetoric

Can Nature Improve Technology?

Peter Coates

In his renowned essay "What is an American?" (1782), J. Hector St. John de Crèvecoeur, the Frenchman-turned-American, invited an English settler fresh off the boat to contemplate previous colonists' tremendous accomplishments: "Here he beholds fair cities, substantial villages, extensive fields, an immense country filled with decent houses, good roads, orchards, meadows, and bridges, where an hundred years ago all was wild, woody, and uncultivated!"[1] Though he chose the adjective *uncultivated* rather than *unimproved*, at the core of Crèvecoeur's environmental ideology resided the notion of improvement. Applied alike to the nonhuman world of nature and the mental, moral, and spiritual faculties of humanity, *to improve (on)* signified "to make better"—to render more serviceable, profitable, and productive. Education improved the mind; religion improved morals; and cultivation improved the land.

Standard notions of the improved and unimproved in land and nature were inverted half a century later by an even more famous French commentator on American life. Alexis de Tocqueville, who visited the northwest frontier in 1831, lamented that nature unimproved held little interest for the American, who, "living in the wilds . . . only prizes the works of man. He will gladly send you off to see a road, a bridge. . . . But that one should appreciate great trees and the beauties of solitude—that possibility completely passes him by."[2]

The roads and bridges that Americans were so keen to show off to visiting Europeans more interested in seeing what was scarce back home—the works of nature gloriously unimproved—were in the vanguard of "opening up" the country. These so-called internal improvements, which also included canals, canalized river channels, and railroads, rendered unexploited natural resources more accessible and connected consumers with the commodities manufactured from these raw materials.[3]

The starting point for this examination of the relationship between technology, nature, and the ethos of improvement will be the powerful, readily demonstrable conviction deeply rooted in Western culture that technology produces a better nature—a nature, that is, that better serves human needs. Whether technology can improve nature is not, however, this essay's central theme. I want to stand this initial proposition on its head. Rather than posing the question, "can technology improve nature?" let us inquire, "can nature improve technology?" When we expand our understanding of technology to include nature's ingenuity as well as our own, we push the discussion into less familiar territory. In doing so, we shall examine themes such as remediation of damage from some technologies through other technologies, bioremediation, and biomimetics, as well as concepts such as living technology, the naturalization of technology, and nature as engineer.

Improving Nature with Technology

Though references to "conquering," "taming," and "subduing" nature were rife in the times of Crèvecoeur and Tocqueville, Americans rarely set out to destroy willfully. In his poem "Song of the Broad-Axe" (1856), Walt Whitman chose an ancient tool as the arch-symbol of the American enterprise. In American hands, the blood of victims that had defined its Old World role as a perverted instrument of murder was "washed entirely away," and the axe was restored to its worthy status as an instrument for making the world more habitable.[4]

As Whitman intimated, the desire to improve the human condition by reordering nature's previous arrangements with technological assistance goes to the very heart of what it means to be human. From ancient Rome to the American frontier, the concept of improvement was as pervasive as the ethos of conquest. Like classical Roman thinkers who advocated the creation of a "second world within the world of nature," promoters of agrarian settlement in the American West regarded the God-given natural world as unfinished. Human completion of nature through technological means was the realization of a divinely ordained plan. The Reverend E. R. Dille proclaimed in an address to the California State Agricultural Society in 1885: "The farmer is given the high honor of finishing and improving the Creator's work. He made the sea, and the mountains, and the heavens as he would have them, complete at first. But the earth, with its animal and vegetable tribes, he only made in the rough and left man to put on the finishing."[5]

Euro-Americans also pursued the "finishing" of nature through biotic (living) technologies in the form of non-native animals and plants. Acclimatizers released carp from Germany in American rivers and fish native to the east coast of the United States in California.[6] And although southern California may have struck colonists as a terrestrial Eden, they nonetheless insisted that even Eden could be perfected with the help of exotic trees (notably the eucalyptus) and the applied science of plant breeding. The surname of Luther Burbank, the late nineteenth-century Californian "plant developer" who wanted to nurture bigger prunes and smaller peas, eventually became a synonym for improvement, featuring in *Webster's Dictionary* as a transitive verb: "Burbank, *v. t.* To modify and improve (plants or animals), esp. by selective breeding. . . . Hence, figuratively, to improve . . . by selecting good features and rejecting bad, or by adding good features."[7] Human intervention in the dynamics of evolutionary development to reshape organisms into "tools for human convenience and profit" has continued apace.[8]

Mass enthusiasm for betterment through technological modification of the natural world has been the leitmotif of modernity and is integral to the Western world's dedication to "progress." A few maverick voices questioned the benefits of progress conventionally defined in the mid-nineteenth century. Henry David Thoreau decried much-vaunted recent inventions, with specific reference to the mechanical telegraph, as "improved means to an unimproved end." And the mixed blessings of "improvement" were the central theme of George Perkins Marsh's pioneering conservationist text, *Man and Nature* (1864). Marsh's question could be formulated as, *"how far can technology improve nature?"* His main reason for writing the book was to indicate "the evils resulting from too extensive clearing and cultivation, and other so-called improvements."[9]

Yet the belief in nature's improvability through technology remained pervasive. In the late nineteenth century, the now familiar abstract and generic notion of technology emerged, replacing a sense of the specificity of individual mechanical artifacts that performed particular tasks (encapsulated in notions of the mechanical or practical arts) and assuming the dimensions of a "disembodied" force with its own logic and agenda.[10]

Since the 1960s, though, a reaction against the worship of technology has converted the idiosyncratic skepticism of Thoreau into a more widespread apostasy. A new order of technological threat constituted by substances such as radioactive fallout, synthetic detergents, and chemical fertilizers and pesticides shaped a new environmental consciousness. Loss of faith in our ability to reengineer the world to our advantage defined the "age of ecology." In their desire to impose limits, so-called technological pessimists replaced reification with a vilification that not only questioned the notion of using technology to its full extent but sometimes bordered on a rejection of technology as intrinsically harmful—not only to nonhuman nature but to us as well. Growing appreciation of our connections with the rest of nature brought a realization that environmental and human impacts were inseparable.[11]

Does technology produce "sustainable abundance, or ecological crisis"? Does it produce "more security, or escalating dangers"? These are two fundamental questions that David Nye recently posed. Whether alleged improvements are demonstrable in the sense that they are amenable to measurement or quantification according to agreed criteria—whether technology is *able* to improve nature as distinct from whether *we believe that* technology can improve nature—is another matter entirely. In the case of fish transplantations, a beneficial introduction could be defined as one that flourishes without collateral damage: the new species either coexists with natives or occupies hitherto vacant space. What is clear, though, is that no matter how far-reaching and deplorable their consequences have turned out to be, the outcome of our technological interventions is not so much the "death of nature" as the creation of another (remade) nature.[12]

Remediation through Technology

Over the past fifteen years, a fresh, more discriminating question has been formulated that departs from the traditional environmentalist tendency to characterize technology as a monolithic entity to be treated with extreme caution at best: can new technologies actually help mitigate or repair damage that older technologies have inflicted?

In 2005 the Sierra Club's magazine inquired, "can technology save the planet?" This was a call for a revolution in the environmentalist's standard attitude: rather than being invariably treated as the planet's enemy, our technical ingenuity ought to be embraced as a potential friend of the earth. The lead article portrayed a series of technological pioneers applying technological solutions to ecological problems. To cite just two examples: the use of robotic dogs mounted on all-terrain wheels to sniff out toxic compounds among the debris in landfill sites; and, to protect forests in the developing world, the manufacture of charcoal for cooking fuel from the fiber of sugar cane after the juice has been squeezed out.[13]

Advanced technology of this kind already delivers results. In the early 1990s, the Tennessee Valley Authority (TVA) began to redesign the blades of its hydroelectric turbines to increase the amount of dissolved oxygen in the water passing through its dams. Meanwhile, injections of gaseous oxygen from perforated hoses boost oxygen levels at the bottom of the slack tailwaters of the impoundments behind TVA dams. Fish are returning to the Tennessee River watershed, as are their predators (blue herons, otters, mink, and weasels).[14]

The new meaning that the notion environmental improvement has acquired since the 1960s is embodied in the Environmental Improvement Award, offered by the Chemical Institute of Canada's Environmental Division since 1975 to the project that makes the most important contribution to the prevention, treatment, or remediation of a pollution problem.[15] Had environmental improvement awards been available a century earlier, they would have been bestowed on a railroad, a dam, or a new strain of drought-resistant wheat that pushed the farming frontier into previously inauspicious regions. Environmental improvement is now defined as enlightened technological solutions to ecological problems attributable to an unenlightened use of technology. An early example was the modifications to the original design for the trans-Alaska oil pipeline to counteract the danger of melting permafrost and to facilitate wildlife migration.[16]

Innovative pipeline design undeniably mitigated potentially negative environmental impacts. Yet improvements directly if unintentionally attributable to technological advance are also claimed. Offshore oil-drilling operations challenge the received wisdom on the ecological dangers of advanced industrial technologies. Since the blowout on a platform six miles off the coast of Santa Barbara, California, in January 1969, offshore rigs have been perceived primarily as a hazard from an environmental standpoint. But they offer an opportunity too.

The average life span of an offshore rig is just a quarter of a century, and most of California's remaining twenty-seven platforms will become obsolete between 2010 and 2025. In the Gulf of Mexico, three hundred oil and gas platforms have already reached the end of their productive lives. Dismantling a decommissioned structure is not the only option, however. A rig can receive a new lease on life as an artificial reef. A rig in the Gulf of Mexico attracts marine organisms from the day it is installed, as the structure provides a firm substrate on a muddy marine bed. Rigs also provide more habitat levels and opportunities for shelter than do natural reefs, whose height rarely exceeds twenty-five feet. Rig removal, which requires explosives, kills or mortally wounds large numbers of fish and deprives encrusted invertebrate communities of their homes. Retention, by contrast, conserves the new habitats and their occupants.

To the delight of recreational divers and sport fishermen, various Gulf states have preserved more than a hundred rigs since 1979. Exxon relocated an experimental production system from Louisiana waters to a designated artificial reef site off Florida in 1979. A platform was toppled in its original location for the first time in Florida in 1987. ("Reefing" entails beheading, i.e., removing the portion of the structure that protrudes above the water, slicing the underwater part into horizontal portions, and rearranging these slices on the seabed next to the anchored base.)

"Rig to reef" proposals have proved more controversial in Southern California, with conversion plans abandoned in 1974 and 1988 and all redundant platforms removed to date. The California Artificial Reef Enhancement Program (CARE), which seeks to promote public awareness of their "vibrant" ecological value, emphasizes that rigs attract crustaceans, starfish, and rockfish. "Higher densities of rockfishes and lingcod at platforms compared to natural outcrops, particularly of larger fishes," contends a team of marine biologists, "support the hypothesis that platforms act as de facto marine refuges."[17]

Yet California environmentalists tend to be much less enthusiastic than their Gulf-states counterparts. Reefing, they argue, allows oil companies to shirk their legal (and expensive) obligation to restore the status quo on the seabed. Environmentalists also contest the assumption that rigs boost species populations overall, protesting that they simply encourage organisms to relocate and do not increase reproductive rates.[18] Environmental enhancement or obstacle? The terminology and technology may have changed, but the question remains the same as it was for Tocqueville, Thoreau, and Marsh: does technology improve nature or detract from its existing value?

In the examples discussed so far, the role of natural organisms themselves has been essentially reactive. Various technological applications have dictated the terms within which nature operates. Any benefits accrued have been indirect and unforeseen. Yet there are also examples of nature enjoying a more assertive relationship with technology. Certain biotic entities can proactively help solve problems our technologies create. The most familiar instance of the applied science of bioremediation is the mobilization of single-celled microorganisms to accelerate the natural breakdown of toxic substances such as crude oil. This microbiological degradation, which works through nutrient enrichment, encourages the rapid proliferation of naturally occurring microorganisms that break down hydrocarbons. Bacterial agents have been deployed against oil spills to supplement methods of chemical dispersal and mechanical cleanup since the late 1960s. Oil-eating bugs were marshaled in the cleanup campaign after the *Torrey Canyon* ran aground off the Cornish coast in 1967, yet their remedial role did not attract large-scale American public attention until after the *Exxon Valdez* tanker spill in Alaskan waters in the spring of 1989. Oil-industry representatives and Environmental Protection Agency (EPA) officials propounded the virtues of degrading hydrocarbons nature's way. "It's natural. It's promising," remarked EPA chief William Reilly. The company responsible for oil operations in Alaska, Alyeska, hailed them as "Alyeska's Microscopic Clean-up Crew."[19]

Improving Technology with Nature

The oil-eating microbes unleashed onto Prince William Sound's soiled beaches in the wake of the *Exxon Valdez* are a prime example of organisms designated as "living technology" and "eco-machines." These terms may initially sound odd and implausible at best, downright wrongheaded and heretical at worst. They may also strike environmental historians and environmentalists accustomed to thinking in terms of the well-worn oppositional categories of nature and technology as oxymoronic. As Edmund Russell has observed, "In the minds of many historians, technology has consisted of machines.... Hiding behind this view is an assumption about the relationship between technology and nature: technology replaced or modified nature, but nature was not technology." Biologists working in the realm of biomechanics are entirely comfortable with the view that nature has its own technologies. Animal mechanics, a concept dating back to the late nineteenth century, seeks to integrate physics, mathematics, and engineering with biology by examining the physics and engineering of animal organisms. How do birds fly, ants crawl, badgers burrow, fleas jump, frogs hop, fish swim, and water beetles walk on water?[20]

Whereas learning about nature's workings from technology and engineering is the primary goal of animal mechanics, a more recent perspective encourages engineering and technology to learn from nature. In the vanguard of biomimetics (also referred to as biomimicry) is a group of thinkers and inventors who call themselves "bioneers." According to Paul Hawken, "bioneering" deploys "biological know-how that defies the conventional idea of what technology is."[21]

Biotechnology, for most people, denotes the modification of plant and animal genes. Yet biomimicry promotes itself as an alternative version of biotechnology that represents the antithesis of manipulative genetic engineering. The mission of the "true biotechnologies" is to tap into the nearly 4 billion years of so-called evolutionary wisdom contained within "nature's operating instructions."[22] Many claims for nature's inspirational "teachings" from this store of accumulated knowledge appear in the literature of biomimicry. Nostril hairs inspired the invention of dust filters; lobster claws inspired pliers; bird beaks inspired tweezers; thorny hedges inspired barbed wire; the scallop shell inspired the corrugated surface; the octopus inspired the suction cup; a dried cocklebur inspired Velcro; and the dandelion seed inspired the parachute. We are also instructed that nature's technologies are often superior to their human-manufactured analogues. The spider's silk is three times stronger, ounce for ounce, than steel, and five times more shatterproof than Kevlar, used for bullet-proof vests. Mother-of-pearl, or nacre, the substance that oysters and other molluscs secrete to form the insides of their shells, is tougher than any ceramic currently available and more tenacious than superglue.

"Green" engineering is all the rage. Researchers trying to improve ultrasound equipment investigate bat echolocation, while those developing superior navigational systems for robots study how bat ears and rodent whiskers are designed to avoid collisions. The latest in wind-farm technology is almost ready to leave the laboratory for sea trials. The spinning propellers of enormous wind turbines that will be moored to the seabed like oil platforms mimic the aerodynamic quality of sycamore seeds. As their British architect, Martin Pawlyn, explains, "We are going back to first principles, taking our inspiration from nature." He also indicates that the design principle for a new type of desalination plant made of glass and steel that uses evaporation and condensation mechanisms to generate fresh water is the Namibian fog-basking beetle, whose desert-survival strategy is built on a shell that serves as a surface to condense moisture. On foggy nights the usually diurnal beetle crawls out of the sand and clambers to the crest of a dune, where maximum condensation occurs. Lowering its head and raising its derriere, the beetle confronts the fog-bearing wind with its back, on which the moisture condenses before trickling down to its mouth.[23]

Meanwhile, the new "bionic" concept car of the Chrysler Group (formerly, Daimler Chrysler) adopts the shape and lightweight body structure of the boxfish. According to the manufacturers, "The boxfish has its home in the coral reef, lagoons and seaweed of the tropical seas, where it has a great deal in common with cars in many respects. It needs to conserve its strength and move with the least possible consumption of energy, which requires powerful muscles and a streamlined shape. It must withstand high pressures and protect its body during collisions, which requires a rigid outer skin. And it needs to move in confined spaces in its search for food."[24]

The larger process at work in these current examples of nature-inspired design might be described as the naturalization of technology. As a counterpoint to the notion of "industrializing organisms" (the title of a collection of essays about redesigning plants and animals),25 a series of essays on the inspirational and instructional qualities of bat ears, sycamore seeds, fog-basking beetles, and boxfish might be entitled *Naturalizing Technologies* (subtitled, perhaps, *No Improvement in Nature Needed, or Possible*). What a "bioneer" regards as the improvement of technology through nature's "wisdom," however, other, more conventional inventors and engineers might simply see as further evidence of a process as old as human existence, namely, the ongoing improvement on nature's provision and capabilities (regardless of whether the prime source of inspiration is humans or other-than-human nature). A more important issue for present purposes is how inventors and engineers have learned from the nature beyond ourselves.

Various aspects of how technologies are naturalized by learning from nature require more rigorous investigation than they currently receive in biomimics' writings. These include the character of the inspirational role that advocates of biomimicry claim for natural substances and processes; the relationship between "naturfact" and "artifact"; and the attitude toward nonhuman nature of the nature-inspired inventor.26 I shall examine four well-known American and European inventions: barbed wire, the conservatory (hothouse), the airplane, and Velcro.

Barbed Wire

Though they are often conflated, deriving inspiration from nature is not the same as copying nature. Tweezers and tongs are direct applications of the mechanism of the bird beak, just as fins for diving and snorkeling are directly copied from the webbed feet of frogs and geese (hence the name *frogman*). Likewise, the hook-and-barb head of a harpoon mimics a bee stinger. Inspiration is a less tangible, more indirect form of influence than copying.

In his classic history of the nineteenth-century settlement of the Great Plains, in which technological innovation looms large, Walter Prescott Webb acknowledged that a natural entity "furnished the basic idea" for one of the key technologies of agrarian colonization: barbed wire. Writing over half a century later, George Basalla was less equivocal, asserting that barbed wire "was not created by men who happened to twist and cut wire in a particular fashion. It originated in a deliberate attempt to copy an organic form that functioned effectively as a deterrent to livestock." In Basalla's study of how technology evolves, barbed wire exemplifies an important general phenomenon: the transformation of the "naturfact" into the artifact. The "naturfact" in this instance was a bush.[27]

When cattle were set to graze on the largely treeless western plains after the Civil War, farmers sought a substitute for the stone walls and wooden fences that enclosed livestock in the eastern United States. Osage orange, a thorny bush native to Texas and Arkansas, supplied an initial, stopgap solution. The next step was the literal combination of natural and human technologies ("naturfact" and artifact). One of the first to patent and purvey barbed wire, Jacob Haish, owned a lumberyard in Illinois that sold Osage orange. As he recalled in 1881, "It was in my mind [between 1857 and 1872] to plant osage-orange seed, and when of suitable growth cut and weave it into plain wire and board fences, using the thorns as safeguard against the encroachments of stock." Haish soon realized, though, that fastening wood to wire did not work. He explained how his thinking evolved: "First was the osage-orange idea, next attachments of metal to wood . . . later I saw wire married to wire and no divorce."[28]

Barbed wire's other inventors also appear to have been directly inspired by nature's example. With reference to Michael Kelly's patent application (1868), an early historian of barbed wire, Charles Washburn, commented that the inventor "prefers to call his structure a thorn wire or thorny wire, having in mind the analogy between his wire and the thorn hedge."[29] This close bond between nature and technology was reflected in the name of Kelly's manufacturing firm, the Thorn Wire Hedge Company.

Basalla's belief that technological innovation entails more evolution than discovery meshes with the biomimetic conviction that nature has its own instructive technologies on which many human inventions are based. An invention that emerged from a different social context and physical environment, in Victorian Britain, further supports this contention.

The Conservatory

In a conservatory (hothouse in contemporary parlance) at the Duke of Devonshire's Chatsworth House, in Derbyshire, in November 1849, the first British flowering of the giant Amazonian lily, *Victoria regia*, was unveiled to an enraptured public. Yet visitors were not just smitten by the floral beauty of the lily, with its blooms over a foot wide and smelling of pineapple; they were equally impressed by the size of its leaves (pads). Four to six feet wide, with a rim measuring two to five inches and a purple-red underside, these leaves were routinely compared to enormous floating tea trays. Their strength also at-

Can Nature Improve Technology? | 53

Miss Annie Paxton standing on a gigantic lily leaf at Chatsworth. (From "The Gigantic Water-Lily [Victoria Regia], at Chatsworth," *Illustrated London News* 15 [17 November 1849]: 328)

tracted attention. "Springing from the end of the petiole or leaf-stalk," John Fisk Allen explained, "where it joins the leaf are eight main ribs which diverge constantly into numerous lesser ones; and these diverging in all directions, strengthened also by arched or curved cross ribs or ties, afford the requisite firmness and support, and exhibit a truly wonderful mechanism." The *Gardener's Chronicle* noted "the extreme buoyancy of its large succulent foliage, occasioned by the presence of large air-cells in the thick ribs which cover like network the under surface, much aided no doubt by its large surface, and the deep pit-like recesses formed between the interlacing veins."[30] The recesses between the largest ribs (veins) were two inches deep at the center of the leaf. The rigidity of the main radiating ribs was combined with and complemented by the flexibility of cross ribs.

Witnesses marveled most, though, at the load-bearing capacity of these robust leaves. Dressed as a fairy, the eight-year-old, forty-two-pound daughter of Chatsworth's head gardener, Joseph Paxton, took her stand on a four-foot-wide pad and almost stole the show. This stunt was often repeated, frequently with portly men. The head gardener of the botanical garden at Ghent, in Belgium, was anxious to test the lily pad's limits. He installed a man on a pad in an attempt to sink it, but to no avail. No less than 760 pounds of bricks—equivalent to five men of average weight—ultimately did the trick.[31]

What inspired the design for the conservatory that made it possible for this prized tropical plant to flourish during an English winter? Paxton attributed it to the lily pad's tremendous resilience. He revolutionized building design by deploying glass panes in association with cast-iron and wrought-iron frames according to a ridge-and-furrow pattern. The Chatsworth conservatory, in turn, served as Paxton's prototype for the considerably larger Crystal Palace, which graced the Great Exhibition at the World's Fair of 1851 in London.[32]

Paxton took a lily pad to London to illustrate his talk at the Royal Society of Arts in November 1850 about the genesis of his Crystal Palace design for the Great Exhibition. As the Royal Society's transactions subsequently explained, the leaf's underside represented a "beautiful example of natural engineering in the cantilevers which radiate from the centre" and the cross girders between the ribs. "You will observe," Paxton explained to his audience, "that nature was the engineer in this case. If you will examine this [leaf], and compare it with the drawings and models, you will perceive that nature has provided it with longitudinal and transverse girders and supporters, on the same principle that I, borrowing from it, have adopted in this building."[33] Prior to the instruction that Paxton received from the giant lily, it would have been impossible to imagine constructing a building as large as the Crystal Palace without raising massive walls.

As the *Cottage Gardener* pointed out, Robert Schomburgk, the German botanist generally credited with discovering *Victoria regia* in British Guiana in 1837, "little thought that in giving the world this magnificent aquatic he gave it as well the germ of that Palace which now crowns the hill of Sydenham."[34] The general acceptance by that time that the link in this instance between naturfact and artifact, however unlikely, was direct does not mean that Paxton's reliance on what he called natural engineering can be equated with nature-inspired design as understood by today's biomimicists. Does Paxton's conservatory in fact reveal more about time-honored ideas of controlling nature than about novel notions of deference to and reverence for nature, not to mention nature's agency? After all, the lily pad's mechanistic properties brought it to Paxton's attention.

Nor was Paxton alone in being attuned to the giant lily's design potential. On seeing his first *Victoria regia* in its native habitat in 1849, the British plant hunter Richard Spruce was initially confounded: "The aspect of the Victoria in its native waters is so new and extraordinary that I am at a loss to what to compare it." Groping for analogies, his first reference point was a series of floating tea trays. On further inspection and reflection, however, he was struck by an analogy with a familiar feature of industrial England. "A leaf, turned up," he noted, "suggests some strange fabric of cast iron, just taken out of the furnace, its colour, and the enormous ribs with which it is strengthened increasing the similarity." What this demonstrates for the historian Margaret Flanders Darby is how predisposed Victorian Englishmen were, no matter how far from home, "to find technology while looking for nature."[35]

The danger of falling into the trap of anachronism when evaluating the spirit in which naturfacts have been converted into artifacts in the past is highlighted again by an invention that Janine Benyus, a prominent current authority on biomimetics, hails as "the last really famous biomimetic invention": the airplane. The avian influence on aviation provides a further opportunity to dissect what is actually meant by the stock phrases of biomimicry, such as *nature's inspiration, nature's lessons,* and *doing it nature's way,* and references to copying, imitating, and emulating nature.[36]

The Airplane

Benyus does not elaborate, but she was clearly referring to the Wright brothers. The most immediate human influence on the Wrights was the French aviation pioneer Louis Pierre Mouillard. Mouillard's book *The Empire of the Air (L'empire de l'air)*, published in 1881, which summarized thirty years of avian research through observation, bore the subtitle *An Essay upon Ornithology as Related to Flight*. A pamphlet extract from this study, which the Smithsonian Institution published in English translation in 1893, was among the materials that the Smithsonian sent to Wilbur Wright following his request for information in the spring of 1899. Mouillard exhorted his fellow humans to cast their gaze upwards: "O! blind humanity! open thine eyes and thou shalt see millions of birds and myriads of insects cleaving the atmosphere. All these creatures are whirling through the air without the slightest float; many of them are gliding therein, without losing height, hour after hour, on pulseless wings without fatigue: and after beholding this demonstration, given by the source of all knowledge, thou wilt acknowledge that Aviation is the path to be followed."[37]

Both Icarus and Leonardo da Vinci were convinced that the secret of flight lay in creating a mechanical version of a *flapping* wing. So was the late nineteenth-century German aviation pioneer Otto Karl Lilienthal, who believed in the exact emulation of a natural form. As Lilienthal's translator, Adolf William Isenthal, explained in his preface to the English edition of Lilienthal's book on bird flight and aviation, Lilienthal set out to "emulate the great model, so constantly exhibited to us by nature, viz. the bird." And Lilienthal himself concluded that "the only possibility of attaining efficient human flight lies in the exact imitation of bird-flight with regard to the aerodynamic conditions."[38]

Mouillard's approach, however, was one crucial step removed from literal biomimicry. He spent much of his life in North Africa, where his intensive scrutiny of birds like vultures, which soar, swoop, and glide for hours on outstretched, motionless wings, convinced him of the possibility, through the "skilful use of the power of the wind, and no other force," of flight in fixed-wing gliders:

> All my life shall I remember the first flight which I saw of the Gyps fulvus, the great tawny vultures of Africa. I was so impressed that all day long I could think of nothing else; and indeed there was good cause, for it was a practical, perfect demonstration of all my preconceived theories concerning the possibilities of artificial flight in a wind.
> ... He has evolved a peculiar mode of flight; he sails and spends no force, he never hurries, he uses the wind instead of his muscles, and the wing flap occasionally seen is meant to limber up rather than to hasten through the air.[39]

A fellow aviation pioneer, Octave Chanute, who reviewed *Empire of the Air*, emphasized the critical role of bird observation in shaping Mouillard's conclusion that given the right combination of wing, tail, and body shape and

"Secret of Aerial Flight Wrested from the Birds," *St. Louis Republic*, 11 March 1906. (From Wright Brothers Scrapbook 1, p. 7, Wright Brothers Collection, Dayton and Montgomery County [Ohio] Public Library)

The Wright brothers observing hovering buzzards. (From unidentified newspaper clipping, Scrapbook 1, p. 8, Wright Brothers Collection, Dayton and Montgomery County [Ohio] Public Library)

weight, a bird could "sail indefinitely upon the wind without further flapping his wings."[40]

Mouillard's observations and conclusions inspired the Wright brothers but did not directly influence the details of their research. As Orville Wright reflected in a letter of 1912 discussing the role of the "elder statesmen of aviation," "Mouillard was an enthusiastic observer of the flight of birds, but I do not think there was anything in his work that contributed scientifically to the solution of the problem. Mouillard was a poet rather than a scientist, and it was [sic] to the charm of his writings in enthusing others in the work that the world owes a debt of gratitude." The Wrights watched vultures at The Pinnacles, a popular picnic spot south of their hometown of Dayton, Ohio, and frequently reported on the flight of raptors and gulls at Kill Devil Hills and Kitty Hawk, North Carolina. But Wilbur attributed his crucial insight into lateral balance—the simultaneous movement of the wing tips, rather than the shifting of its weight, to turn the body—to observations of pigeons, most likely in the summer of 1899.[41]

John Evangelist Walsh is not convinced, though, of the overriding importance of unmediated observation. He claims that Wilbur's vital insight "resulted not from a direct sighting of the wing movements—no eye could follow such slight and rapid fluttering on so small a scale—but by a species of deduction that implies many hours of observation."[42] Wilbur corroborated what he had deduced from his observations of pigeons with the results of his acute study of another bird, writing to Chanute that

> my observation of the flight of buzzards leads me to believe that they regain their lateral balance, when partly overturned by a gust of wind, by a torsion of the tips of the wings. If the rear edge of the right wing tip is twisted upward and the left downward the bird becomes an animated windmill and instantly begins to turn, a line from its head to its tail being the axis. It thus regains its level even if thrown on its beam ends, so to speak, as I have frequently seen them. I think the bird also in general retains its lateral equilibrium, partly by presenting its two wings at different angles to the wind, and partly by drawing in one wing, thus reducing its area.[43]

As well as alluding to attempts to "imitate" buzzard flight at Kitty Hawk in 1901, Wilbur employed the phrase "if you would copy nature" with regard to the size and proportions of airplanes. In 1902 he advised an aircraft engineer that "this agrees with what we find in birds, as the tip to tip measurement [of the wing] is never less than six times the longitude of wing from front to rear, and in the sea birds, which live on the wing, the tip to tip spread is sometimes twenty times the fore and aft dimension. *If you would copy nature* your surfaces should have a lateral breadth not less than six times the length fore and aft, nor more than twenty times."[44]

Were the so-called birdmen students of animal mechanics rather than incipient biomimics? Were they inventing rather than imitating or discovering? Like their fellow aviation pioneers, the Wrights used birds to confirm their theories rather than as direct transmitters of evolutionary wisdom. This particular lesson from nature did not take the form of instant revelation, let alone direct emulation. Wilbur recollected that "the conclusion did not come until some unspecified time after the sighting" and surmised that more conventional influences—books—may have been of greater importance in the short term.[45]

However, newspapermen loved the idea that birds were providing humans with flight instruction. A typical headline in 1906 read, "Secret of Aerial Flight Wrested from the Birds." And the early twentieth-century public found the label "birdman" and its connotations irresistible. Writing in *Col-*

lier's magazine in 1909, a medical doctor propounded the rather fanciful notion that unlike the overwhelming mass of people who were descended from fish, pilots could trace their ancestry back to birds.[46]

That the Wright brothers frequently and intensively observed birds is unquestionable. Yet as Orville recalled in 1941, "I cannot think of any part bird flight had in the development of human flight excepting as an inspiration. Although we intently watched birds fly in a hope of learning something from them I cannot think of anything that was first learned in that way. After we had thought out certain principles, we then watched the birds to see whether it [sic] used the same principles." He drew an analogy with learning how a magician performs tricks: until you know what to look for, there are certain things you simply do not see.[47]

The biologist Steven Vogel believes that genuine instances of "profitable" copying are rare. He identifies fewer than a dozen "acceptable cases of bio-emulation." In most cases of alleged bioemulation, he concludes, "nature may have played some role, but inflating her contribution demeans splendid engineering achievements." Characterizing the desire to see nature as instructor as "bucolic romanticism," Vogel aims to "ruffle our tendency to view nature as the gold standard for design and as a great source of technological breakthroughs." In his view, crediting nature with an influential role in the shaping of technology means "taking a pretty dim view of human creativity."[48]

Whether it follows that the celebration of nature's inventiveness necessarily entails the disparagement of our own ingenuity and resourcefulness is unclear. Still, Vogel's approach to the relationship between nature and technological invention has a good deal to commend it. Many purported examples of copying nature involve a questionable understanding of causality. There is an enormous difference between identifying commonalities—between a winged seed and a helicopter, for example—and demonstrating a clear and direct route from naturfact to artifact. My final historical example is an invention for which that route is remarkably clear and direct (even Vogel agrees).[49]

The Velcro Brand Hook-and-Loop Invention

The zipper's nemesis, the never-jamming method that revolutionized systems of fastening and closure first industrially and then domestically in the early 1980s, was devised by a Swiss electrical engineer. George de Mestral was a keen inventor. He designed a toy plane at the age of twelve, and after the Velcro brand fastening systems, he patented a hygrometer (a device for measuring air humidity) and an asparagus peeler. In 1999 he was inducted into the National Inventors Hall of Fame, in Akron, Ohio. Yet Mestral was an equally enthusiastic wildfowl hunter. He often returned home with the dried flowers of a mountain thistle known as the cocklebur stuck to his woolen hunting pants and to the coat of his gun dog. After a hunt in the Jura Mountains in the summer of 1941, he decided to analyze one of these infernal burs under a microscope. The seedpod's exterior was completely covered with tiny but very strong hooks that attached themselves to the equally small soft loops that made up the surface of his pants and his dog's fur.[50]

Over years of trial and error, Mestral strove to convert naturfact into artifact. He eventually discovered that if nylon (itself a relatively new substance) was sewn under infrared light, practically unbreakable hooks developed on its surface. Described on the Velcro Company's UK Web site (which displays an image of a dried cocklebur) as "the most natural fastener,"[51] Velcro brand hook-and-loop tape has three hundred hooks and loops per square inch. Patented in 1951, the name derives from the French words *velour* (velvet) and *crochet* (hook). Velcro was not an instant hit, though the aerospace industry adopted it in the early 1960s as a fastener for clumsy spacesuits, which frustrated fumbling fingers. Though hook-and-loop tape now seems ubiquitous—many grateful children have been freed from the chore of tying shoelaces—new frontiers remain to be conquered. (In the 2004 movie *Garden State* a layabout explains how he became an overnight millionaire by inventing silent hook and loop and selling the patent to the U.S. government, presumably for military clothing.)

Conclusion

Technology is conventionally defined as the means whereby we incorporate the nonhuman world of nature into the human world and convert it to our use. Technology acts as an intermediary connecting people with nature.[52] Yet, as the examples above demonstrate, the nonhuman world of nature can also mediate between people and technology. When nature is incorporated into culture, culture is also incorporated into nature. Similarly, while technology has modified nature, nature has also modified human society by shaping our technology. An "improved" form of technology refers, increasingly, to a technological instrument or method rooted in nature's "design."

Studying what has been dubbed natural engineering, bioemulation, biomimicry, and the imitation and copying of nature alters traditional understandings of biotechnology and the notion of technology transfer as well. Looking at how nature can improve technology also expands our appreciation of nature's role as model and agent. Treating the natural world as a repository of moral lessons dates back at least to Aesop's fables. But a technological perspective offers an updated version of the "book of nature" by approaching the natural world as a source of inspiration and guidance for inventors and engineers, as well as poets, philosophers, and sociologists.

Not least, the examination of barbed wire, the conservatory, aerodynamics, and Velcro illuminates the spirit in which the improvement of technology through naturalization is pursued. A leading current practitioner, Julian Vincent, defines biomimetics as the "technological outcome of the act of borrowing or stealing ideas from nature." Yet "stealing" hardly conveys the notion of "respectful imitation," which is Benyus's definition of biomimicry.[53] Does the approach of biomimetic inventors constitute a radical departure from our traditional instrumental view of the natural world as a collection of inert entities to be manipulated and transformed at will? Or does it amount to a recycled, green-lacquered version of the mechanistic, Cartesian world-view, which regards all creatures as automata, and the reductive, Baconian view of nature as a storehouse of useful materials to serve human needs? Alternatively, are we reading too much into an unreflective choice of verb?

The ethic of appropriation for the sake of bettering human life, taking from nature something that has been withheld but is rightfully ours, is by no means absent from the thinking of aviation's pioneers. Otto Lilienthal referred to nature as "model" and wrote of the need to "imitate" her wonderful ways. And there is no reason to question his respect for nature's "ingenuity." Yet Lilienthal also spoke the time-honored language of mastery over nature. When he reflected that nature had "denied" humans the power of flight, "deplore[d] the inability of man to indulge in voluntary flight," and urged "victory" over the air, he propounded the need to extend human powers.[54]

Those who suspect that using the terminology of theft to characterize the attitude of biomimetic inventors toward nonhuman wisdom is more than just an unfortunate choice of words will be unimpressed by two advertisements in the *New Yorker* in 2006. These advertisements suggest that our usual relationship between technology and nature's improvement has been revolutionized. The emergence of a more modest and flexible technological outlook eminently capable of celebrating nature's nonimprovability is evoked in a two-page spread on 27 November 2006 announcing the role of Samsung Heavy Industries in developing "ocean-friendly," liquid-natural-gas-powered, double-hulled supertankers (the *Exxon Valdez*, which ran aground and ruptured, was single-hulled). The main caption accompanying the picture of a massive tanker gliding past a breaching gray whale reflects, "Sometimes contributing to the world means leaving it just as it is." "Preserving the world itself," elucidates the small print, is "one of the many ways in which we're committed to making this a better world."

An advertisement for Lexus automobiles in the 12 June 2006 issue shifts from the redefinition and part renunciation of improvement through technology to an implicit engagement with the tenets of biomimicry. This time the behemoth is nature itself (in the previous advertisement the tanker towers over the whale). Featuring a gargantuan octopus and a tiny saloon car, the thrust of the advertisement is the vehicle's ability to grip the road "even if Mother Nature throws something nasty your way." The caption reads: "Lexus All-Wheel Drive Technology. There's No Comparison. *At Least Not on Land.*"

Those alert to words such as *steal* will probably dismiss these advertisements as just another example of "green-washing," the reduction by corporate image-makers of profound ideas to glib slogans and lofty rhetoric. Others will regard the transfer of such sentiments from the environmentalist periphery to the corporate mainstream as much more encouraging. Why? Because how this ship and this automobile are presented for our visual consumption signals a substantial softening of the arrogance of a humanist technological outlook that conceives of improvement in purely human terms and approaches the nature-technology nexus largely from the vantage point of what technology does to nature. Technology, according to James Williams, is "an instrument of culture" that binds us to the rest of the natural world as much as it separates us from it.[55] Yet technology is also an instrument of nature.

Notes

1. J. Hector St. John de Crèvecoeur [Michel-Guillaume Jean de Crèvecoeur], *Letters from an American Farmer and Sketches of Eighteenth-Century America* (1782; reprint, London: Penguin, 1981), 66.
2. Alexis de Tocqueville, "A Fortnight in the Wilds" (1860), appendix 4 in *Democracy in America*, trans. G. Lawrence, ed. J. P. Mayer and M. Lerner, vol. 2 (New York: Harper & Row, 1966), 968, 974.
3. John Lauritz Larson, *Internal Improvement: National Public Works and the Promise of Popular Government in the Early United States* (Chapel Hill: University of North Carolina Press, 2001), 1–8.
4. Walt Whitman, *Leaves of Grass and Other Writings*, ed. Michael Moon (New York: Norton, 2001), 155–64.
5. Cicero, *De Natura Deorum*, quoted in *Traces on the Rhodian Shore: Nature and Culture in Western Thought from Ancient Times to the End of the Eighteenth Century*, by Clarence J. Glacken (Berkeley and Los Angeles: University of California Press, 1967), 145; E. R. Dille, quoted in Linda Nash, "Finishing Nature: Harmonizing Bodies and Environments in Late-Nineteenth-Century California," *Environmental History* 8 (January 2003): 39.

6. Jerry C. Towle, "Authored Ecosystems: Livingston Stone and the Transformation of California Fisheries," *Environmental History* 5 (January 2000): 54–74.
7. *Webster's New International Dictionary*, 2d ed., s.v. "burbank." In *A Culture of Improvement: Technology and the Western Millennium* (Cambridge, MA: MIT Press, 2007), Robert Friedel contends that the driving force behind technological change, fueled by the deep-seated conviction that we can do things better, is the pursuit of improvement.
8. Deborah Fitzgerald's phrase appears on the back cover of Susan R. Schrepfer and Philip Scranton, eds., *Industrializing Organisms: Introducing Evolutionary History* (New York: Routledge, 2004).
9. Merritt Roe Smith, "Technology, Industrialization, and the Idea of Progress in America," in *Responsible Science: The Impact of Technology on Society*, ed. Kevin B. Byrne (New York: Harper & Row, 1986), 1–20; Leo Marx, "Does Improved Technology Mean Progress?" *Technology Review* 90 (January 1985): 33–41, 71; Henry David Thoreau, *Walden* (1854; reprint, New York: New American Library, 1980), 40; George Perkins Marsh, quoted in David Lowenthal, introduction to Marsh's *Man and Nature; Or, Physical Geography as Modified by Human Action*, ed., with a new introduction, by David Lowenthal (Seattle: University of Washington Press, 2003), xxii.
10. Leo Marx, "The Idea of 'Technology' and Postmodern Pessimism," in *Technology, Pessimism, and Postmodernism*, ed. Yaron Ezrahi, Everett Mendelsohn, and Howard Segal (Boston: Kluwer Academic, 1994), 14–25.
11. George Basalla, *The Evolution of Technology* (New York: Cambridge University Press, 1988), 133; Everett Mendelsohn, "The Politics of Pessimism: Science and Technology, Circa 1968," in Ezrahi, Mendelsohn, and Segal, *Technology, Pessimism, and Postmodernism*, 151–74.
12. David E. Nye, *Technology Matters: Questions to Live With* (Cambridge, MA: MIT Press, 2006), 87–108, 161–84; Towle, "Authored Ecosystems," 71. The phrase *death of nature* is from Carolyn Merchant, *The Death of Nature: Women, Ecology, and the Scientific Revolution* (San Francisco: Harper & Row, 1980).
13. Bruce Sterling, "Can Technology Save the Planet?" *Sierra* 90 (July–August 2005): 32–35; Dashka Slater, "Earth's Innovators," ibid., 36–41, 71.
14. See http://www.tva.com/environment/water/rri_triblist.htm#hiwassee; http://www.tva.gov/environment/reports/envreport/fishfriendly.htm; and http://www.tva.gov/environment/water/rri_results.htm.
15. See http://www.cheminst.ca/index.cfm?ci_id=1964&la_id=1.
16. Peter A. Coates, *The Trans-Alaska Pipeline Controversy: Technology, Conservation, and the Frontier* (Bethlehem, PA: Lehigh University Press, 1991), 254–61.
17. See http://www.calreefs.org/; and Milton S. Love, Donna M. Schroeder, and Mary M. Nishimoto, *The Ecological Role of Oil and Gas Production Platforms and Natural Outcrops on Fishes in Southern and Central California: A Synthesis of Information* (Seattle: Biological Resources Division, U.S. Geological Survey, U.S. Department of the Interior, 2003), vii.
18. Anna Davison, "Two New Studies Highlight Oil Rigs' Importance to Fish" and "Environmentalists Would Rather See Rigs Removed," *Santa Barbara News-Press*, 30 June 2006, A1, A12.
19. Carol Hunter, "Natural Solutions: Organic Answers to Toxic Questions," *Berkeley Science Review* 3 (Fall 2003): 34–41; Craig Medred, "EPA Chief Likes Bugs that Eat Oil," *Anchorage Daily News*, 6 August 1989, B3.
20. John Todd, "Living Technologies: Wedding Human Ingenuity to the Wisdom of the Wild," in *Nature's Operating Instructions: The True Biotechnologies*, ed. Kenny Ausubel (San Francisco: Sierra Club Books, 2004), 17–32; Randall von Wedel, "Bioremediation: Waste Equals

Food," ibid., 42–49; Edmund P. Russell, "Introduction: The Garden in the Machine; Toward an Evolutionary History of Technology," in Schrepfer and Scranton, *Industrializing Organisms*, 1; R. McNeill Alexander, *Animal Mechanics* (London: Sidgwick & Jackson, 1968); S. A. Wainwright et al., *Mechanical Design in Organisms* (London: E. Arnold, 1976).

21. Paul Hawken, foreword to Ausubel, *Nature's Operating Instructions*, ix.
22. Kenny Ausubel, introduction to Ausubel, *Nature's Operating Instructions*, xiv.
23. Yoseph Bar-Cohen, *Biomimetics: Biologically Inspired Technologies* (London: CRC, 2005), 25; John Vidal, "Engineers Race to Steal Nature's Secrets," *Guardian* (London), 29 August 2006, 13.
24. Sandrine Ceurstemont, "Nature-Inspired Design," 31 August 2006, http://www.firstscience.com/site/editor/0154_ramblings_31082006.asp; Steven Cole Smith, "Fish Story," http://www.edmunds.com/insideline/do/Columns/articleId=106111/subsubtypeId=218.
25. Schrepfer and Scranton, *Industrializing Organisms*.
26. For "naturfact" and "artifact," see Basalla, *Evolution of Technology*.
27. Walter Prescott Webb, *The Great Plains* (Boston: Ginn, 1931), 298; Basalla, *Evolution of Technology*, 50–55.
28. Jacob Haish, quoted in Charles G. Washburn, "History of the Manufacture of Barbed Wire Fencing" (c. 1920), quoted in Webb, *Great Plains*, 300.
29. Washburn, "History of the Manufacture of Barbed Wire Fencing," quoted in Webb, *Great Plains*, 301.
30. John Fisk Allen, *Victoria Regia: or, the Great Water Lily of America. With a Brief Account of its Discovery and Introduction into Cultivation* (Boston: Dutton & Wentworth, 1854), http://www.victoria-adventure.org/victoria_images/allen_sharp/victoria_regia.html; *Gardener's Chronicle* 41 (12 October 1850): 645.
31. Samuel Lockwood, "The Victoria Regia: Historical Reminiscences," *Transactions of the Wisconsin State Horticultural Society* 15 (1885): 144–45.
32. "The 'Crystal Palace,'" *Times* (London), 14 November 1850, 5.
33. *Transactions, Royal Society of Arts* 57 (1850–51): 1–57, quoted in ibid., 4; Joseph Paxton, quoted in "The Grand Industrial Exhibition for 1851," *Illustrated London News*, 17 August 1850, 141.
34. I.H.C., "History and Cultivation of the Victoria Regia," *Cottage Gardener and Country Gentleman's Companion* 19 (1858): 83.
35. Richard Spruce, quoted in ibid.; Margaret Flanders Darby, "Joseph Paxton's Water Lily," in *Bourgeois and Aristocratic Cultural Encounters in Garden Art, 1550–1850*, ed. Michel Conan (Washington, DC: Dumbarton Oaks Research Library and Collection, 2002), 18.
36. See Janine M. Benyus, *Biomimicry: Innovation Inspired by Nature* (New York: William Morrow, 1997), 8, 9.
37. Wilbur Wright, "What Mouillard Did," *Aero Club of America Bulletin* 1 (April 1912): 3–4; *The Papers of Wilbur and Orville Wright, Including the Chanute-Wright Letters and Other Papers of Octave Chanute*, ed. Marvin W. McFarland, 2 vols. (New York: McGraw-Hill, 1953), 1:4–5; Louis Pierre Mouillard, quoted in Wright, "What Mouillard Did," 3.
38. Otto Karl W. Lilienthal, *Birdflight as the Basis of Aviation*, trans. Adolf William Isenthal (London: Longmans, Green, 1911 [1889]), vi, vii, 128.
39. Louis Pierre Mouillard, *Empire of the Air: An Essay upon Ornithology as Related to Flight*, trans. Octave Chanute, Smithsonian Institution Pamphlet 903, in *Annual Report for 1892* (Washington, DC, 1893), quoted in Wright, "What Mouillard Did," 3. The original publication was *L'empire de l'air: Essai d'ornithologie appliquée a l'aviation* (Paris: Libraire de L'Académie de Médecine, 1881).

40. Octave Chanute, "Progress in Flying Machines," pt. 8, "Aeroplanes," *Railroad and Engineering Journal*, January 1893, http://invention.psychology.msstate.edu/i/Chanute/library/Prog_Aero_Jan1893.html.
41. Orville Wright to Henry Woodhouse, 9 December 1912, in *Miracle at Kitty Hawk: The Letters of Wilbur and Orville Wright*, ed. Fred C. Kelly (1951; reprint, New York: Da Capo, 2002), 397; excerpt from Wilbur's Notebook, 1900, in ibid., 36–37.
42. John Evangelist Walsh, *First Flight: The Untold Story of the Wright Brothers* (London: Allen & Unwin, 1976), 36.
43. Wilbur Wright to Octave Chanute, 13 May 1900, in Kelly, *Miracle at Kitty Hawk*, 23. Fifteen bird observations appear in Notebook A (September–October 1900), published in McFarland, *Papers of Wilbur and Orville Wright*, 1:34–37.
44. McFarland, *Papers of Wilbur and Orville Wright*, 2:1126; Wilbur Wright to George A. Spratt, 23 January 1902, in Kelly, *Miracle at Kitty Hawk*, 60–61, emphasis added.
45. Walsh, *First Flight*, 258–59. The books were J. Bell Pettigrew, *Animal Locomotion* (1874) and Étienne Jules Marey, *Animal Mechanism: A Treatise on Aerial Locomotion* (1890).
46. *St. Louis Republic*, 11 March 1906, 10; Joseph J. Corn, "Making Flying 'Thinkable': Women Pilots and the Selling of Aviation, 1927–1940," *American Quarterly* 31 (Autumn 1979): 559, wherein the medical doctor Vance Thompson's views are noted.
47. McFarland, *Papers of Wilbur and Orville Wright*, 1:60–487 passim, 2:1126–27; Orville Wright to J. Horace Lytle, 27 December 1941, in ibid., 2:1168–69.
48. Steven Vogel, *Cats' Paws and Catapults: Mechanical Worlds of Nature and People* (New York: Norton, 1998), 249, 10, 250–51, 256.
49. Ibid., 268–70.
50. Allyn Freeman and Bob Golden, *Why Didn't I Think of That? Bizarre Origins of Ingenious Inventions We Couldn't Live Without* (New York: John Wiley, 1997), 99; Reinhard Budde, "The Story of Velcro," *Physics World* 8, no. 1 (1995): 22; Bar-Cohen, *Biomimetics*, 9; "This Month in Physics History," *APS (American Physical Society) News* 13 (February 2004): 2.
51. See http://www.velcro.co.uk/cms/History.6.0.html.
52. Theodore R. Schatzki, "Nature and Technology in History," *History and Theory* 42 (December 2003): 82, 92.
53. Julian F. V. Vincent, "Stealing Ideas from Nature," in *Deployable Structures*, ed. S. Pellegrino (Vienna: Springer-Verlag, 2001), 51; Benyus, *Biomimicry*, 2.
54. Lilienthal, *Birdflight as the Basis of Aviation*, 94, 2, 1, 139.
55. James Williams, "Understanding the Place of Humans in Nature," in this volume.